INSTRUCTOR'S MANUAL
WITH TEST ITEM FILE

ESSENTIALS OF MANAGEMENT

INSTRUCTOR'S MANUAL WITH TEST ITEM FILE
Craig Hollingshead
Texas A&M University - Kingsville
with Lisamarie Brassini Caby

ESSENTIALS OF MANAGEMENT

Gary Dessler

Prentice Hall
Upper Saddle River, New Jersey 07458

Acquisitions editor: *Stephanie K. Johnson*
Managing Editor: *Melissa Steffens*
Associate editor: *Hersch Doby*
Project editor: *Theresa Festa*
Manufacturer: *Technical Communication Services*

Printed in the United States of America

10 9 8 7 6 5 4 3 2 1

ISBN 0-13-012898-8

Prentice-Hall International (UK) Limited, *London*
Prentice-Hall of Australia Pty. Limited, *Sydney*
Prentice-Hall Canada Inc., *Toronto*
Prentice-Hall Hispanoamericana, S.A., *Mexico*
Prentice-Hall of India Private Limited, *New Delhi*
Prentice-Hall of Japan, Inc., *Tokyo*
Prentice-Hall (Singapore) Pte Ltd
Editora Prentice-Hall do Brasil, Ltda., *Rio de Janeiro*

CONTENTS

Chapter 1

MANAGING IN THE 21ST CENTURY

INTRODUCTION

This chapter provides an overview of the field of management as we enter the 21st century. It explains what managers do, how they accomplish work through people, and how they must respond to important environmental changes to achieve organizational effectiveness.

LECTURE OUTLINE

What's Ahead?

Managing in an Era of Change
- Technological Innovation
- Globalization
- Deregulation
- Changing Political Systems
- Demographics and the New Global Workforce
- Human Capital: A Shift to Service and Knowledge Work

What Managers Do
- The Management Process
- Types of Managers

The People Side of Management
- The People Side of Management: An Example
- Building Three Core Processes
- How Managers Get the Job Done -- Through People
- Rearranging the Management Process
- Do You Want to Be a Manager ?

The Future Is Now: Snapshots of the Modern Organization
- ABB Asea Brown Boveri
- Saturn Corporation
- AT&T
- Tomorrow's Management Today
- On to the Future

Summary

Case: What is the Future of Trend Micro ?
You Be the Consultant: Knitting Music is UP and Running

ANNOTATED OUTLINE

What's Ahead ?

Paul "Kinko" Orfalea began the Kinko Copy Center business at California State University in Santa Barbara by borrowing $5,000 to provide students with 24-hour access to word processors and copiers for their reports. He built a loyal following and subsequently expanded nationwide. His business has remained successful by adapting services offered to changes in the environment, including the product market.

Managing in an Era of Change

This section explains important environmental trends that impact on the practice of management. Managers face rapid change and uncertainty as they try to manage their organizations successfully into the future.

Notes:

Technological Innovation. Information highways, the Internet, microprocessors, and automated factories are proliferating as technological innovations are changing the way companies compete. Organizations that have adapted well to technological change and used new technology to their advantage have had a significant advantage over those that have not.

Notes:

Globalization. U. S. organizations have been extending sales and manufacturing to new markets abroad to grow the business and try to earn a higher rate of return for investors. The value of U. S. imports/exports grew from 9.4 percent of GNP in 1960 to almost 23 percent in the 1990s. Globalization of markets and manufacturing has vastly increased international competition and put pressure on managers in many organizations to improve productivity, quality, and organizational performance.

Notes:

Deregulation. Removal of the comfortable protection of governmental regulations in industries such as airlines have created significantly more competitive environments in which to do business. The increased competitive pressures have required managers to make changes in their organizations to adapt.

Notes:

Changing Political Systems. Important political changes such as the fall of communism have triggered an explosive opening of new markets in countries such as Russia. Investment opportunities are enormous, but the burgeoning demand for goods and services has created an increase in competition among the countries attempting to acquire market share in the newly created political systems that have evolved.

Notes:

Demographics and the New Global Workforce. Workforce diversity is proliferating. During the 1990s, the white labor force will have grown less than 15 percent, the black labor force will have grown by about 29 percent, and the Hispanic by more than 74 percent. Asian and other nonwhite races will account for more than 11 percent of the growth in the labor force during the decade. These changes are requiring managers to make changes in the way they operate. Diversity training and assimilation of the new minority and multicultural workers will be very important to the success of organizations. Equal opportunity protection laws are being vigorously enforced and will require the continued attention of managers.

Notes:

Human Capital: A Shift to Service and Knowledge Work. The U.S. human capital investment rose from 42 percent of total productive investment to about 58 percent between 1959 and 1990. During the same time, the investment in physical capital as a percentage of total productive investment dropped from about 58 percent to about 45 percent. The investment has been shifted because of the growing service economy and changes in manufacturing-intensive jobs to knowledge-intensive, high-tech jobs in such industries as aerospace, computers, and telecommunications. Innovation also has demanded more highly skilled employees. New products must be developed and marketed continuously to keep up with changes in demand. Knowledge workers often resist the old command and control management model and require new management systems with more emphasis on the behavioral side.

Notes:

What Managers Do

Organization: An organization consists of people with formally assigned roles who must work together to achieve stated goals. Management:. A manager is someone who plans, organizes, leads, and controls the people and the work of the organization in such a way that the organization achieves its goals.

Notes:

Management Process. The management process consists of planning, organizing, leading, and controlling. Planning is setting goals, deciding on courses of action, developing rules and procedures, developing plans, and forecasting. Organizing involves identifying jobs to be done, hiring employees, and establishing an organization with authority and a chain of command to coordinate the work of subordinates. Leading is influencing people to get the job accomplished, developing company culture, and managing conflict and communication. Controlling is setting standards, comparing performance to standards, and taking corrective action as required.

Notes:

Types of Managers. Executives are top managers in the organization. **Middle managers** are responsible for managing functions in the organization such as sales or production. First-line managers, or supervisors, supervise the employees who are providing services or producing products.

Notes:

The People Side of Management

Managerial Roles. Henry Mintzberg identified several managerial roles including figurehead, leader, liaison, spokesperson, and negotiator. The figurehead role includes performing ceremonial tasks. The leader role includes creating a motivational climate for employees. The liaison role includes maintaining contact with people outside their own departments or outside the organization. The spokesperson role involves serving as communicator for the organization, keeping others informed of developments. The negotiator role involves negotiating settlement of issues as required.

Notes:

The People Side of Management: An Example. Chairman Lawrence A. Bossidy is an example of the type of manager who is successful in organizations, today. He has people skills, and when selecting managers, he looks for those who have a gift for working with and turning on employees. Managers cannot rely on traditional authority and vertical hierarchy to accomplish thing in organizations, today. Instead, they need to know how to persuade, coach, and build consensus. They need to add value by brokering with people, not presiding over empires.

Notes:

Building Three Core Processes. Managers must focus on the entrepreneurial, competence-building, and renewal processes to get people to focus attention on creating a responsive and change-oriented company. Entrepreneurial processes refer to the externally-oriented, opportunity-seeking attitudes that motivate employees to run their operations as if they own them. Competence-building processes require managers to devote efforts to the creation of an environment that permits employee competence to flourish by encouraging employees to take on more responsibility and providing education and training to assist. The renewal process requires managers to guard against complacency, develop the habit of continuously questioning work processes, and decide if the status quo should be disrupted.

Notes:

How Managers Get the Job Done – Through People. Managers get their jobs done by working with and interacting with people. They work with subordinates and accept that there may be some hostility and resentment as they assess the strengths and weaknesses of subordinates and deal with competition and conflict within their organizations.

Notes:

Rearranging the Management Process. Managers must recognize that the leading or behavioral side of what they do is not just a step in the management process--it is an integral part of almost everything the manager does.

Notes:

Do You Want to Be a Manager? Personality is an important determinant of career choice. Almost all managers fit either the social orientation or the enterprising orientation personality types as identified by John Holland. People with a social orientation are attracted to careers that involve working with people in a helpful or facilitative way. Those with an enterprising orientation like working with people in a supervisory or persuasive way, in order to achieve some goal. Interests and competencies help determine how successful a person might be in managing others. Analytical, interpersonal, and emotional competence are important in achieving a management position of high responsibility.

Notes:

The Future is Now: Snapshots of the Modern Organization

ABB Asea Brown Boveri is an electrical equipment manufacturer that has reorganized itself to compete in the fast-moving global market by organizing around mini-units, empowering its workers, flattening its hierarchy, and eliminating central staff.

Notes:

Saturn Corporation is a team-based organization where work on the shop floor is organized around relatively autonomous work teams that are responsible for completing a task without traditional supervisors. The team members supervise themselves and do their own hiring, budget control, and quality monitoring.

Notes:

AT&T has undergone tremendous change in the past decade. It reorganized the company by eliminating 140,000 jobs, empowered employees, and instilled new business values throughout the organization. These new values included respect for individuals, dedication to helping customers, integrity, innovation, and teamwork.

Notes:

Tomorrow's Management Today. Here is a summary of what management experts believe "the new management" will look like:

- The average company will be smaller, employing fewer people.
- The traditional organizational structure will become more team-based and boundaryless.
- Employees will be empowered to make more decisions.
- Flatter organizations will be the norm.
- Work itself--on the factory floor, and in the office--will be organized around teams and processes rather than specialized functions.
- The bases of power will change.
- The new organization will be knowledge based.
- The new company will stress vision and values.
- Managers must be change agents.
- Leadership will be more important.

Notes:

On to the Future. Figure 1.3 includes discussion of the fundamental changes facing managers who will have to be more responsive, smaller, flatter, and oriented toward adding value through people in the 21st century. Managers must use planning, organizing, leading, and controlling to make their firms more responsive.

Notes:

Summary

Notes:

Case: What is the Future of Trend Micro?

1. Trend Micro is global and innovative, taking advantage of skilled employees at lower salaries in other countries. It makes products and licenses them in locations that provide it with competitive advantage.

2. Issues confronting U.S. companies include the need to use the best expertise from around the world in solving problems and how to improve mobility in producing and distributing products globally.

3. U.S. organizations with higher salaries and wages must reduce costs where feasible and achieve higher productivity and quality to help capture and hold market share. U.S. firms must develop and maintain the same type of mobility to globally manufacture and distribute.

4. The future for Trend Micro appears to be very bright, as long as it remains innovative and at the forefront in developing and marketing computer virus prevention products. They will need to continue attracting and retaining the best trained experts in the field.

You Be the Consultant: Knitting Music is Up and Running

1. List of trends for which Michael and his colleagues will have to plan:

- Team-based, boundaryless organizations
- Employee empowerment
- Flatter organizations
- Work-teams rather than specialized functions
- Changing bases of power
- Knowledge-based organizations
- Stress on vision and values
- Managers as change agents
- Emphasis on leadership
- Need to blend people and technology more effectively into a better socio-technical work system

2. Students can use Internet resources such as Yahoo to develop a list of competitors in New York City for the Knitting Factory Club. See the online version of this case at www.prenhall.com/desslermgmt

End of Chapter Questions for Chapter 1

Multiple Choice

1. Dozens of banks have been forced to merge to survive, and many U.S. airlines (Eastern, Braniff, USAir, and Pan Am) have either gone out of business or have been forced to merge. Which of the following statements BEST describes what companies that have gone through "merger mania" have attempted to accomplish with their mergers?

a. drive competitors out of business
b. raise tariffs for foreign competitors.
c. increase their marketing and advertising budgets
d. drive up efficiency by squeezing more productivity from acquisitions
e. increase bottom-line share price of stock

Answer: (d) Difficulty: (2)

2. One of the results of technological innovations such as information highways, the Internet, microprocessors, and automated factories is that the number of U.S. ______________ and ______________ have dramatically increased in the 1990s.

a. automobiles and television sets
b. patents and trademarks
c. schools and universities
d. manufacturing companies and television sets
e. exports and fast food restaurants

Answer: (b) Difficulty: (2)

3. Technological innovations in the 1990s are changing the way companies ________

a. control costs.
b. stimulate demand.
c. compete.
d. pay taxes.
e. calculate profit.

Answer: (c) Difficulty: (2)

4. Inter-Design of Ohio (because of mounting pressure from Target stores) has introduced electronic data interchange. Today, their orders arrive via modem, straight into company computers. This is an example of the impact of ______________________

a. information technology.
b. direct manufacturing.
c. speedy service.
d. competitive retailing.
e. depersonalization of the retail environment.

Answer: (a) Difficulty: (1)

5. ________________ merges communications systems with computers.

a. Data acquisition
b. Information technology
c. Tele-purchasing
d. Satellite ordering systems
e. Point-of-sale ordering systems

Answer: (b) Difficulty: (1)

6. _________ is the tendency for firms to extend their sales or manufacturing to new markets abroad.

a. Exporting
b. Importing
c. Tele-marketing
d. Globalization
e. Uni-trading

Answer: (d) Difficulty: (1)

7. U. S. exports are reaching new markets, manufacturing facilities are being placed in markets around the globe, international competition is at an all-time high, and there is a growing integration of the world's economy. All of the above are results of ____________________

a. exporting.
b. importing.
c. tele-marketing.
d. globalization.
e. uni-trading.

Answer: (d) Difficulty: (2)

8. The Toyota Camry (what many would claim is obviously a Japanese car) is produced in Georgetown, Kentucky, and contains almost 80 percent U.S.-made parts. At the same time, the General Motors Pontiac LeMans (an obvious U.S. car) actually contains almost two-thirds foreign-made parts. The above are examples (or the result) of the impact of _______________

a. exporting.
b. importing.
c. shrinking markets.
d. globalization.
e. uni-trading.

Answer: (d) Difficulty: (2)

9. All of the following have a dramatic impact on the way companies operate and the way they are managed in the 1990s EXCEPT ____________________

a. technological innovation.
b. globalization.
c. the rise of communism in Europe.
d. deregulation.
e. changing demographics and the impact on the workforce.

Answer: (c) Difficulty: (1)

10. One of the trends in the 1990s is a changing workforce. Which of the following racial groups has seen the largest growth in the 90s (estimated to be 74 percent)?

a. White
b. Black
c. Hispanic
d. Asian
e. Indian (sub-continent)

Answer: (c) Difficulty: (3)

11. The results of demographic changes in the workforce include all the following changes EXCEPT ____________________

a. special diversity-management programs are being implemented.
b. a hiring freeze on certain racial groups is occurring.
c. equal opportunity protection laws are being aggressively applied.
d. organizations are implementing new training programs for minorities.
e. organizations are instituting assistance programs that attempt to assimilate minorities and those of multicultural backgrounds into the workforce.

Answer: (b) Difficulty: (2)

12. Human capital is changing the way the workplace operates. There is a shift away from physical capital to human capital resources. All of the following are examples of components of human capital (the firm's workers) EXCEPT ____________________

a. strength.
b. knowledge.
c. training.
d. skills.
e. expertise.

Answer: (a) Difficulty: (2)

13. The growing emphasis on education and human capital reflects several trends in the business environment. One is the growing importance of service work in today's society. Over ________________ of the U.S. workforce is now involved in producing services, not things.

a. thirty-five percent
b. forty-five percent
c. one-half
d. two-thirds
e. seventy-five percent

Answer: (d) Difficulty: (2)

14. Which of the following types of job categories will most probably put a bigger premium on worker education and knowledge than do traditional jobs?

a. light manufacturing
b. automated manufacturing
c. services
d. government work
e. housing construction

Answer: (c) Difficulty: (2)

15. According to recent surveys, manufacturing-intensive jobs (such as steel, auto, rubber, and textile industries) are being replaced by ____________________ high-tech manufacturing jobs (such as aerospace, computers, and telecommunications).

a. labor-intensive
b. automation-intensive
c. parts-intensive
d. union-intensive
e. knowledge-intensive

Answer: (e) Difficulty: (3)

16. Today, companies are demanding innovation as a means to meet and surpass competition. Firms are creating new products, new services, and discovering new ways of turning out goods as never before. This means that companies are relying more on the employee's ________________ than ever before. The trend is expected to continue.

a. physical labor
b. conditioning
c. brain power
d. network contacts
e. bargaining skills

Answer: (c) Difficulty: (1)

17. Take an underperforming organization and install a skilled manager, and the manager can soon have the enterprise humming like a well-oiled machine. Take a successful business and put a less-than-competent manager in charge and suddenly the organization will begin a downward spiral to ruin. These statements illustrate the ____________

a. supply and demand effect.
b. human resource training effect.
c. Delphi effect.
d. importance of organizational design.
e. "manager" effect.

Answer: (e) Difficulty: (2)

18. "People with formally assigned roles who must work together to achieve stated goals" best described a(n) ________________

a. formal hierarchy.
b. organization.
c. corporation.
d. team.
e. partnership.

Answer: (b) Difficulty: (3)

19. A ____________ is someone who plans, organizes, leads, and controls the people and the work of the organization in such a way that the organization achieves its goals.

a. leader
b. captain of industry
c. controller
d. manager
e. developer

Answer: (d) Difficulty: (1)

20. The purpose of a manager is to insure that an organization ______________

a. achieves its goals.
b. makes money.
c. makes sales.
d. pays its bills.
e. avoids lawsuits.

Answer: (a) Difficulty: (1)

21. Management refers collectively to the managers of an organization and ____________

a. how they succeed.
b. the study of what managers do.
c. how to make profits.
d. how to control people.
e. how to lead.

Answer: (b) Difficulty: (3)

22. Management writers traditionally refer to the manager's four basic functions of planning, organizing, leading, and controlling as the ________________

a. organizational concept.
b. business hierarchy.
c. management process.
d. management development.
e. professionalism.

Answer: (c) Difficulty: (2)

23. All of the following would be considered to be part of the traditional functions of the management process EXCEPT ________________

a. persuading.
b. planning.
c. controlling.
d. leading.
e. organizing.

Answer: (a) Difficulty: (1)

24. Part of the process of ____________ is setting goals, deciding on courses of action, and developing rules and procedures.

a. persuading
b. planning
c. controlling
d. leading
e. organizing

Answer: (b) Difficulty: (2)

25. ______________ entails identifying jobs to be done, hiring people to do them, and coordinating the work effort.

a. Persuading
b. Planning
c. Controlling
d. Leading
e. Organizing

Answer: (e) Difficulty: (2)

26. ________________ means influencing other people to get the job done, maintaining morale, molding company culture, and managing conflicts and communication.

a. Persuading
b. Planning
c. Controlling
d. Leading
e. Organizing

Answer: (d) Difficulty: (2)

27. Part of the process of ____________ is setting standards and comparing actual performance with these standards.

a. persuading
b. planning
c. controlling
d. leading
e. organizing

Answer: (c) Difficulty: (2)

28. Which of the following management processes would most likely be responsible for the forecasting (that is, predicting or projecting what the future holds for the firm) function?

a. persuading
b. planning
c. controlling
d. leading
e. organizing

Answer: (b) Difficulty: (3)

29. Which of the following management processes would most likely be responsible for establishing a chain of command?

a. persuading
b. planning
c. controlling
d. leading
e. organizing

Answer: (e) Difficulty: (3)

30. One of the simple ways to distinguish between the different types of managers is to characterize them based on their ____________________

a. organizational level, salary, and communication skills.
b. organizational level, position, and functional title.
c. leadership potential, position, and communication skills.
d. network contacts, salary, and years with the company.
e. international experience, salary, and communication skills.

Answer: (b) Difficulty: (3)

31. According to data presented in the text, the best example of an executive would be a __________

a. Production Manager.
b. Sales Manager.
c. Regional Sales Manager.
d. Chief Financial Officer.
e. Chief Bookkeeper.

Answer: (d) Difficulty: (2)

32. The best example of a first-line manager would be a ____________________

a. Vice President of Human Resources.
b. Production Supervisor.
c. Sales Manager.
d. Finance Manager.
e. Human Resource Manager.

Answer: (b) Difficulty: (3)

33. From the list provided below, determine which of the listed activities or management processes are most likely to receive the most attention and time by top managers.

a. planning and budgeting
b. planning and setting goals
c. working on special projects
d. directing and controlling those who work on special projects
e. translation of goals into specific projects

Answer: (b) Difficulty: (2)

34. From the list provided below, determine which of the listed activities or management processes are most likely to receive the most attention and time by middle managers.

a. planning and budgeting
b. planning and setting goals
c. working on a project
d. directing and controlling those who work on special projects
e. translation of goals into specific projects

Answer: (e) Difficulty: (3)

35. From the list provided below, determine which of the listed activities or management processes are most likely to receive the most attention and time by first-line supervisors.

a. planning and budgeting
b. planning and setting goals
c. working on a project
d. directing and controlling those who work on special projects
e. translation of goals into specific projects

Answer: (d) Difficulty: (2)

36. Over 20 years ago, Henry Mintzberg conducted a study of what managers actually do. He found that managers basically wore different hats as they went from task to task. When a sales manager takes an important client to lunch or performs duties of a ceremonial nature, the manager is performing the ______________ role.

a. figurehead
b. leader
c. liaison
d. spokesperson
e. negotiator

Answer: (a) Difficulty: (2)

37. Over 20 years ago, Henry Mintzberg conducted a study of what managers actually do. He found that managers basically wore different hats as they went from task to task. When managers attempt to motivate and encourage their employees, they are performing a _______________ role.

a. figurehead
b. leader
c. liaison
d. spokesperson
e. negotiator

Answer: (b) Difficulty: (1)

38. Over 20 years ago, Henry Mintzberg conducted a study of what managers actually do. He found that managers basically wore different hats as they went from task to task. If a vice-president of sales meets with the vice-president for finance (to make sure that a new customer has the credit required to place an order) or acts as a "middleman" between several departments, then that manager is performing the ____________ role.

a. figurehead
b. leader
c. liaison
d. spokesperson
e. negotiator

Answer: (c) Difficulty: (1)

39. If the president of an organization makes a speech to local county commissioners and attempts to persuade them to grant the organization permission to build a new plant on some unused land, then the president is performing the _______________ role. This role was one of five roles first described by Henry Mintzberg in his studies of what managers actually do.

a. figurehead
b. leader
c. liaison
d. spokesperson
e. negotiator

Answer: (d) Difficulty: (2)

40. If a first-line supervisor attempts to resolve a difference between the company and the organization's union representative, then this manager is performing the ___________________ role. This role was one of five roles first described by Henry Mintzberg in his studies of what managers actually do.

a. figurehead
b. leader
c. liaison
d. spokesperson
e. negotiator

Answer: (e) Difficulty: (1)

41. According to Bartlett and Ghoshal, successful managers cultivate processes aimed at getting people in the comp-any to focus their attention on creating change. These processes can be described as the entrepreneurial process, the competence-building process, and the _____________ process.

a. delegation
b. innovation
c. renewal
d. international
e. directional

Answer: (c) Difficulty: (3)

42. According to Bartlett and Ghoshal, the _____________ process refers to the externally-oriented process that attempts to motivate employees to run their operations as if they own them.

a. entrepreneurial
b. competence-building
c. renewal
d. delegation
e. directional

Answer: (a) Difficulty: (1)

43. According to Bartlett and Ghoshal, the _____________ process refers to creating an environment where employees are encouraged to take more responsibility, providing the education and training necessary to build employee self-confidence, allowing the employees to make mistakes without fear of punishment (helping them to learn from mistakes), and fostering teamwork.

a. entrepreneurial
b. competence-building
c. renewal
d. delegation
e. directional

Answer: (b) Difficulty: (2)

44. According to Bartlett and Ghoshal, the _____________ process is designed to challenge a company's strategies and the assumptions behind them. Above all else, employees are encouraged to guard against complacency.

a. entrepreneurial
b. competence-building
c. renewal
d. delegation
e. directional

Answer: (c) Difficulty: (3)

45. Everything a manager does requires leading. Which of the following managerial functions best illustrates the usage of leadership in the example cited below? Example: A manager attempts to get department heads to work together to craft a new strategic plan.

a. planning
b. organizing
c. developing
d. controlling
e. expediting

Answer: (a) Difficulty: (2)

46. Everything a manager does requires leading. Which of the following managerial functions best illustrates the usage of leadership in the example cited below? Example: Encouraging communication across departmental lines.

a. planning
b. organizing
c. developing
d. controlling
e. expediting

Answer: (b) Difficulty: (3)

47. Everything a manager does requires leading. Which of the following managerial functions best illustrates the usage of leadership in the example cited below? Example: Using effective interpersonal communication skills to encourage employees to change the way they do things.

a. planning
b. organizing
c. developing
d. controlling
e. expediting

Answer: (d) Difficulty: (3)

48. Research with John Holland's Vocational Preference Test (VPT) suggests that almost all successful managers fit which of the following personality types or orientations?

a. dictating orientation
b. commanding orientation
c. unity orientation
d. social orientation
e. detail orientation

Answer: (d) Difficulty: (3)

49. With respect to research done by John Holland, it was found that almost all successful managers fit certain personality types or orientations. From his six basic "personal orientations," he found that _______________ people tend to like working with people in a supervisory or persuasive way, in order to achieve some goal.

a. dictating
b. commanding
c. social
d. detail
e. enterprising

Answer: (e) Difficulty: (3)

50. Based on a study of MIT graduates, Schein concluded that managers had a strong managerial competence career anchor. The students that showed a strong motivation to be managers believed they possessed three competence skills that would ensure a successful career in management. Which of the following were the skills that were most important to the test subjects?

a. analytical, mathematical, and computer competence
b. analytical, interpersonal, and scientific competence
c. analytical, interpersonal, and emotional competence
d. emotional, rational, and international competence
e. organizational, legal, and emotional competence

Answer: (c) Difficulty: (3)

51. To fit today's fast-changing, globally competitive, and increasingly high-tech environments, a new breed of organization is arising. Which of the following words best describes the feeling, attitude, or motivation that is the top priority with the aforementioned new breed of organization?

a. responsiveness
b. dictating
c. commanding
d. mechanical
e. formality

Answer: (a) Difficulty: (2)

52. Zurich-based electrical equipment maker ABB (Asea Brown Boveri) is a good example of a firm that disorganized itself to compete in the ever-changing international marketplace. ABB did all of the following to make itself super-responsive EXCEPT ___________________

a. it organized around mini-units.
b. it empowered its workers.
c. it flattened its hierarchy.
d. it eliminated unions.
e. it eliminated central staff.

Answer: (d) Difficulty: (2)

53. One of the most responsive and progressive companies today is the GM subsidiary Saturn Corporation. The primary feature that sets this organization apart from its vastly larger parent is its _____________ organization.

a. creative
b. team-based
c. supervisor-controlled
d. top-management controlled
e. top-down

Answer: (b) Difficulty: (2)

54. AT&T has drastically changed itself in recent years. For example, many of the top managers prefer to be called "coaches" instead of "bosses." Instead of managing in the traditional sense, these new-age managers try to get their "associates" (a term preferable to subordinate) to focus on their customers' needs and respond accordingly. A descriptive term that would best describe this new attitude and style of managing at AT&T would be _______________

a. creative.
b. supervisor-controlled.
c. top-management controlled.
d. product-oriented.
e. empowerment.

Answer: (e) Difficulty: (3)

55. Today, many firms are re-creating themselves. All of the following are estimations of what "the new management" and characteristics of these dynamic firms will look like EXCEPT __________

a. the average company will be larger and will employ more people.
b. the traditional organizational structure will become more team-based.
c. employees will be empowered to make more decisions.
d. flatter organizations will be the norm.
e. the new organization will be knowledge based.

Answer: (a) Difficulty: (2)

True/False and Short Answer

56. Based on the most recent information from the U.S. Patent office, it appears that United States companies have been relying on foreign competition to supply most of the cutting-edge patents over the last several years. The number of U.S. patent and trademark applications is down sharply from a high of 66,000 per year in the 1980s.

Answer: (False) Difficulty: (2)

57. The field of _______________ merges communications systems with computers.

Answer: (information technology) Difficulty: (3)

58. ________________ is the tendency of firms to extend their sales or manufacturing to new markets abroad.

Answer: (Globalization) Difficulty: (3)

59. During the 1990s, the White labor force will have grown less than 15 percent, while the Black labor force will have grown by about 29 percent and the Hispanic by more 74 percent.

Answer: (True) Difficulty: (2)

60. Today, over two-thirds of the workforce is involved in manufacturing goods. This accounts for the rebound in meeting foreign competition in the manufacturing sector.

Answer: (False) Difficulty: (2)

61. An ______________ consists of people with formally assigned roles who must work together to achieve stated goals.

Answer: (organization) Difficulty: (3)

62. A ________________ is someone who plans, organizes, leads, and controls the people and the work of the organization in such a way that the organization achieves its goals.

Answer: (manager) Difficulty: (1)

63. The management process is characterized as being made up of production, pricing, distribution, and promotion of goods and services.

Answer: (False) Difficulty: (2)

64. Planning entails identifying jobs to be done, hiring people to do them, and establishing a chain of command.

Answer: (False) Difficulty: (1)

65. An example of a top manager would be the president of the firm.

Answer: (True) Difficulty: (1)

66. An example of a first-line manager would be the sales manager of the organization.

Answer: (False) Difficulty: (2)

67. If a manager spends part of his or her time performing duties that are of a ceremonial nature, the manager is performing the _______________ role. This role was one those described by Henry Mintzberg over 20 years ago.

Answer: (figurehead) Difficulty: (3)

68. If the president of an organization makes a speech to lobby the local county commissioners for permission to build a new plant on some unused land, the president is performing the coercive role. This role was first described by Mintzberg over 20 years ago as being one of the five roles that managers are called upon to perform.

Answer: (False) Difficulty: (2)

69. ________________ is the study and application of knowledge about how people (as individuals and as groups) act within organizations.

Answer: (Organization behavior) Difficulty: (3)

Essay

70. Managing in an era of change is a difficult task. The purpose of this chapter was to explain what managers do. In order to do that, seven different changes (or areas of potential change) must be recognized. List and briefly identify each of these change areas. Be specific in your characterization.

Answer:

a. Technological innovation: Technological innovations are changing the way companies compete. Information technology (which merges communications systems with computers) is having a dramatic impact on management and management practice.

b. Globalization: Globalization (the tendency of firms to extend their sales or manufacturing to new markets abroad) is dramatically increasing. There are more markets, competition, and new rules of the marketplace.

c. Deregulation: The comfortable protection provided to thousands of businesses around the world by government regulations has been stripped away in country after country.

d. Changing political systems: Communism is on the decline and (with the fall of the U.S.S.R.) capitalism seems to flourishing in some very odd places. New markets have been opened, and millions of new customers have been created.

e. Demographics and the new workforce: There has been a shift from a traditional white workforce in the United States to one of a multicultural background. These changes are changing the way the modern corporation is managed. Many firms are also transferring operations abroad to take advantage of cheap labor in third-world countries

f. More service-oriented jobs: Over two-thirds of the United States workforce is involved in the service sector of the economy. This shift alone is causing management to rethink its manufacturing mentality. Different intellectual skills are required in a service economy.

g. New emphasis on “knowledge-based” work: Training, expertise, and skills must be adjusted to an environment where computer skills are more necessary than strength or manual dexterity. Human capital and brainpower are more precious commodities.

Difficulty: (2)

71. Identify and briefly describe the four basic functions of the management process.

Answer:

a. Planning: Planning is setting goals and deciding on courses of action, developing rules and procedures, developing plans (both for the organization and for those who work in it), and forecasting (that is, predicting or projecting what the future holds for the firm).

b. Organizing: Organizing entails identifying jobs to be done, hiring people to do them, establishing departments, delegating or pushing authority to subordinates, establishing a chain of command, and coordinating the work of the manager's subordinates.

c. Leading: Leading means influencing other people to get the job done, maintaining morale, molding company culture, and managing conflicts and communication.

d. Controlling: Controlling is setting standards (such as sales quotas or quality standards), comparing actual performance with these standards, and then taking corrective actions as required.

Difficulty: (2)

72. Over 20 years ago, Henry Mintzberg conducted a study of what managers actually do. He summarized their activities as being primarily in five areas. Identify each of the roles that managers are expected to perform. Pick one of the roles that you believe to be (in general) the most important and briefly describe why.

Answer:

a. The Figurehead Role: Every manager spends part of his or her time performing some duties of a ceremonial nature.

b. The Leader Role: Every manager must function as a leader, motivating and encouraging his or her employees to do their best.

c. The Liaison Role: Managers spend part of their time in contact with people outside their own departments, essentially acting as the liaison between their departments and the other people within and outside the organization.

d. The Spokesperson Role: The manager is often the spokesperson for his or her organization. This role involves transmitting information and presenting positions. In some instances, the manager must explain or defend the positions of the organization.

e. The Negotiator Role: Mintzberg found that managers also spend a lot of their time negotiating between different factions within and outside the firm.

Students may choose any of the above for their choice as "most important." The instructor should explain standards that might be used to judge the credibility of the answer.

Difficulty: (3)

Chapter 2

MANAGING IN A GLOBAL ENVIRONMENT

INTRODUCTION

This chapter provides an overview of the management function in a global environment. It explains why firms expand operations internationally and explains strategies used to achieve global objectives. It examines how conducting business globally affects managerial planning, organizing, leading, and controlling. Economic, legal, political, sociocultural, and technological factors that influence the decision to expand abroad are examined.

LECTURE OUTLINE

What's Ahead?

Understanding International Business
- The Language of International Business
- Economic Integration and Free Trade
- Why Companies Go Abroad

Types of International Strategies
- Exporting
- Licensing
- Franchising
- Foreign Direct Investment and the Multinational Enterprise
- Joint Ventures
- Wholly Owned Subsidiaries

The Business Team in a Global Economy
- Global Marketing
- Globalization of Production
- Global Staffing

The Global Manager

The Manager in an International Environment
- The Economic Environment
- The Political and Legal Environment
- The Sociocultural Environment
- The Technological Environment

The Process of International Management
- Multinational Planning Issues
- Organizing the Multinational Business
- Leading the International Enterprise
- Issues in Controlling the Multinational Enterprise

Summary

Case: Going Global at Blockbuster Video
You Be the Consultant: Going Global with Michael Dorf and Knitmedia

ANNOTATED OUTLINE

What's Ahead?

Larry Harris, founder of the Pollo Tropical fast-food chain, believes that his best opportunity for market expansion is to move into Mexico and South and Central America. He needs to develop a strategy to succeed in these new, international markets.

Notes:

Understanding International Business

The Language of International Business. Organizations that engage in international trade or investment across national boundaries are involved in international business. International trade is the export or import of goods or services to consumers in another country. International investment is the investment of resources in business activities outside a firm's home country. International management includes performance of the management functions across national borders. An internationally integrated business that is controlled by a parent corporation and owned/managed essentially by nationals of its home country is a multinational corporation. A global corporation operates as if the entire world were a single entity, producing and selling the same product or service in the same way everywhere for global consumption.

Notes:

Economic Integration and Free Trade. Economic integration occurs when two or more nations obtain the advantages of free trade by minimizing trade restrictions between them. The concept of free trade means that all trade barriers among participating countries are removed. In a customs union, member countries dismantle trade barriers between themselves but establish a common trade policy regarding non-members. In a common market, there are no trade barriers between members, a common external trade policy exists, and factors of production move freely between member countries.

Notes:

Why Companies Go Abroad. Sales expansion is usually the primary reason for internationalization efforts, although lower labor costs are sometimes a consideration.

Notes:

Types of International Strategies

Exporting. Exporting is often the initial strategy when a manufacturer decides to expand sales abroad. It involves selling abroad directly to customers or indirectly by foreign sales agents and distributors.

Notes:

Licensing. Licensing is an arrangement where a firm grants a foreign firm the right to use intangible property such as patents, copyrights, manufacturing processes, or trade names for a certain time period, usually in return for a royalty.

Notes:

Franchising. The granting of a right by a parent company to another firm to do business in a prescribed manner is known as franchising. Franchising differs from licensing by requiring longer time periods and much stricter guidelines in running the business.

Notes:

Foreign Direct Investment and the Multinational Enterprise. Firms often need to go beyond exporting, licensing, or franchising in another country and may have to make direct foreign investment, e.g., Toyota at Georgetown, Kentucky.

Notes:

Joint Ventures. The participation of two or more companies jointly in an enterprise in which each party contributes assets, owns the entity to some degree, and shares risk is known as a joint venture. A joint venture is a type of strategic alliance that is more general in nature and could include licensing or franchising agreements.

Notes:

Wholly Owned Subsidiaries. When a subsidiary is owned 100 percent by the foreign firm, it is wholly owned. Wholly owned subsidiaries can have tighter controls over operations but are a relatively costly way to do business.

Notes:

The Business Team in a Global Economy

Global Marketing. Marketing globally has become a necessity. Firms in global industries must globally expand to survive in their domestic markets. Exports have exploded during the 1990s as firms seek to implement global marketing strategies.

Notes:

Globalization of Production. Putting parts of a firm's production process in various locations around the globe to provide manufacturing and supply support for the firm's marketing efforts is globalization of production.

Notes:

Global Staffing. Firms doing business abroad must understand complex employment laws and procedures in other countries. They also encounter complex issues in taxation and compensation when trying to hire people in other countries.

Notes:

The Global Manager.

A manager who views markets and production globally and who seeks higher profits for his or her firm on a global basis is a global manager. The new global manager must be cosmopolitan and comfortable anywhere in the world and in different cultures.

Notes:

The Manager in an International Environment

The Economic Environment. Economic systems range from the market economy of the U.S. to the command economy of the People's Republic of China. A mixed economy includes private ownership and free-market mechanisms along with some organizations owned and managed by the government. Economic development varies significantly, and exchange rates play an important role in profitability.

Notes:

The Political and Legal Environment. Trade controls are governmental influences usually aimed at reducing the competitiveness of imported products or services. Tariffs are governmental taxes levied on goods shipped globally. In some countries, direct payments called subsidies are made to domestic producers. Quotas may be imposed on imports of particular goods. The political and legal environment may range from common law in England to codified law in other countries.

Notes:

The Sociocultural Environment. Different values in various societies are among the most influential cultural differences between countries. Cultural differences among countries may be reflected by power distance, individualism versus collectivism, masculinity versus femininity, or uncertainty avoidance. Power distance is the extent to which less powerful members of society accept and expect that power will be distributed unequally. Individualism or collectivism is measured by whether or not ties between individuals are normally loose or close. Societies that value assertiveness are normally considered to rank high on masculinity. Uncertainty avoidance refers to whether people in the society are uncomfortable with unstructured situations in which unknown or surprising incidents may occur.

Notes:

The Technological Environment. The ability to transfer technology across countries can determine the success of a product. Successful technology transfer depends upon the availability of suitable technology, favorable social and economic conditions for the transfer, and the willingness and ability of the receiving party to use and adapt the technology.

Notes:

The Process of International Management

<u>**Multinational Planning Issues**</u>. Planning in the international environment involves more complex political, legal, and technological issues. Uncertainty from political instability, currency instability, and pressure from national governments such as changing tariff barriers make planning more difficult.

Notes:

<u>**Organizing the Multinational Business**</u>. Companies typically go through several organizational stages as they progress from domestic to multinational enterprises. The stages include domestic, export-oriented, international, and multinational. In the domestic organization, each division handles its own foreign sales. In the export-oriented organization, the firm may use an export-oriented organizational structure where one department coordinates all activities in licensing, contracting, etc. International organizations divide into separate domestic and international divisions. In a multinational firm, each country where the firm does business has its own subsidiary.

Notes:

<u>**Leading the International Enterprise**</u>. Leadership approaches vary significantly by country. Japanese managers value consensus, while in Latin America, bosses are expected to be more autocratic. Companies must ensure leadership approaches are consistent with the environment in which they are conducting business.

Notes:

<u>**Issues in Controlling the Multinational Enterprise**</u>. Geographic distance, language, and cultural barriers can distort communications and make control more difficult in global operations.

Notes:

Summary.

Notes:

Case: Going Global at Blockbuster Video

1. Blockbuster could implement its diversification strategies to respond to the flattening of video rentals. This would involve lower risk and costs than vertical integration. If Simon & Schuster is acquired, their

expertise could be helpful in a diversification strategy. Vertical integration could be considered if Viacom wants to make the investment. The market penetration strategy is a short-run, quick-fix approach not likely to work in the long run. Blockbuster charges more than many rental outlets like drug stores, and they need to offer other advantages, e.g., more movie selection or other products, to offset their higher prices. The geographic expansion idea might work until the other technologies are in place in developing countries.

2. The company would need to research thoroughly the legal-political and cultural difference of any country under consideration. The employment, environmental, and other laws applying to business must be examined to ensure they will not prevent the organization from carrying out the expansion successfully. If the company cannot adapt to the differences, it would need to abandon the plan. The company should not seek to impose its systems on another country and its workers.

3. Blockbuster acquired City Vision PLC in the UK, established a joint venture in Japan, and opened stores in several countries. Blockbuster is a good example of the type of U.S. firm discussed in Chapter where its domestic market growth has slowed and it needs to look globally for expansion and growth. It should continue opening new operations overseas where market research indicates that growth and profit are probable.

4. The person selected should have a thorough understanding of Japanese culture. An ideal candidate would be a Japanese national with an M.B.A. from a U.S. business school. Such a person would know Japanese culture and would be familiar with U.S. business strategies and practices. Compensation issues would be significant, especially the cost-of-living differences, e.g., housing.

You Be the Consultant: Going Global with Michael Dorf and KnitMedia

1. They began by exporting examples of their music by presenting bands on tour in Europe. They established a strategic alliance with the South Bank Centre in London. They are adding more clubs and establishing a global Internet music broadcasting operation in conjunction with MCI. They might want to consider the lower risk of establishing joint ventures in countries where they are less certain of the acceptance of their music and style.

2. The company experienced the impact of exchange rates when Germany was involved in the reunification process. The same problems can be experienced with exchange rates across countries. Technological development via the Internet has opened a new outlet for the firm to broadcast music globally and develop demand for its product in various countries.

3. Planning is increasingly complex and can cause operational problems, e.g., the difficulties with booking the tour in clubs with insufficient facilities. Organizing becomes more difficult, including efforts to coordinate and manage a geographically diverse operation. Control will become more difficult as the firm diversifies its product line and expands globally. Quality control will be more difficult. Leading will become more complex as managers need to provide a sense of direction to an increasingly diverse and geographically widespread workforce.

4. They will face competition from innovative European and other global competitors who perceive an opportunity to enter the market and gain share. They will need to react as competition increases and other firms provide new challenges. Management should monitor the environment continuously to develop strategies for adapting successfully to change.

End of Chapter Questions for Chapter 2

Multiple Choice

1. An ________________ is any firm(s) that engage(s) in international trade or investment.

a. international cartel
b. investment group
c. international business
d. international trade group
e. multinational corporation

Answer: (c) Difficulty: (1)

2. Instead of just focusing on an individual firm's activities, the term ____________ can also refer to general activities such as exporting goods, services, and skills across national boundaries.

a. international cartel
b. investment group
c. international business
d. international trade group
e. multinational corporation

Answer: (c) Difficulty: (2)

3. International trade refers to ____________________

a. actions between buyers and sellers.
b. the export or import of goods or services to consumers in another country.
c. bartering between businesses.
d. an ideology and philosophy of how to do business.
e. a mathematical measurement technique for measuring performance.

Answer: (b) Difficulty: (1)

4. ________________ refers to the investment of resources in business activities outside a firm's home country.

a. International business
b. International trade
c. International investment
d. International management
e. International development

Answer: (c) Difficulty: (3)

5. The term ______________________ is best characterized as being the performance of the management functions of planning, organizing, leading, and controlling across national borders.

a. international business
b. international trade
c. international investment
d. international management
e. international development

Answer: (d) Difficulty: (2)

6. The primary functions included in international management would be characterized as being __________

a. international finance and acquisition.
b. leading and negotiation.
c. barter and acquisition.
d. cultural adaptation and language development.
e. planning, organizing, leading, and controlling.

Answer: (e) Difficulty: (1)

7. Which of the following terms best describes an internationally integrated business that is controlled by a parent corporation, owned and managed essentially by the nationals of its home country?

a. international business
b. international trade group
c. a cartel
d. multinational corporation (MNC)
e. a strategic alliance

Answer: (d) Difficulty: (2)

8. Which of the following examples would be least likely to be a multinational corporation?

a. General Electric
b. General Motors
c. ITT
d. IBM
e. Humana

Answer: (e) Difficulty: (3)

9. Multinational corporations operate manufacturing and marketing facilities in two or more countries. Those operations are coordinated by a parent firm. The owners of that parent firm are most likely to be located in ____________________

a. a neutral country.
b. a tax haven country.
c. the organization's home country.
d. the organization's international market country.
e. a cartel country or region.

Answer: (c) Difficulty: (1)

10. Marketing expert Theodore Levitt contends that the multinational corporation is slowly being displaced by a the ____________________ corporation.

a. global or transnational
b. regionally affiliated
c. bloc or trade group
d. geo-superior
e. non-profit, service

Answer: (a) Difficulty: (2)

11. The chief characteristic of the global corporation is that it operates ____________________

a. in a defined market zone.
b. as if the entire world (or major regions of the world) were a single entity.
c. above national laws.
d. as if it is not accountable to federal governments.
e. as if it is not responsible for taxes in any particular country or region.

Answer: (b) Difficulty: (3)

12. The concept of free trade helps to explain why business today is increasingly international. Free trade means ____________________

a. contracts have been established between companies to do business.
b. regulatory environments make trading fair for all parties.
c. the World Bank monitors and controls enterprise actions.
d. all trade barriers among participating countries are removed.
e. that an idealistic state has been achieved and non-members are prohibited from jeopardizing the achieved balance.

Answer: (d) Difficulty: (2)

13. The classic explanation for the advantages of free trade was presented by _______________ in *The Wealth of Nations.*

a. Max Weber
b. Frederick W. Taylor
c. Adam Smith
d. Karl Marx
e. Herbert Stein

Answer: (c) Difficulty: (3)

14. The basic concept of free trade between and among nations would advocate which of the following terms as a basic construct?

a. specialization of production
b. tariff barriers for protection
c. tight government control and support subsidies
d. generalization (make everything)
e. socialism and a good welfare state

Answer: (a) Difficulty: (2)

15. Through economic integration, which of the following levels best characterizes the phrase "all trade barriers to trade among member countries are removed, so that goods and services are freely traded among the member countries?"

a. customs union
b. cartel
c. regional alliance
d. common market
e. free trade area

Answer: (e) Difficulty: (3)

16. Through economic integration, which of the following levels best characterizes the phrase "members dismantle barriers to trade among themselves while establishing a common trade policy with respect to non-members?"

a. customs union
b. cartel
c. regional alliance
d. common market
e. free trade area

Answer: (a) Difficulty: (2)

17. Through economic integration, which of the following levels best characterizes the phrase "no barriers to trade exist among members, and a common external trade policy is in force; in addition, factors of production, such as labor, capital, and technology, move freely between member countries?"

a. customs union
b. cartel
c. regional alliance
d. common market
e. free trade area

Answer: (d) Difficulty: (2)

18. All of the following are examples of members of the European Union (the largest economic integration) effort on the European continent EXCEPT ____________________

a. United Kingdom.
b. France.
c. Germany.
d. Poland.
e. Austria.

Answer: (d) Difficulty: (2)

19. The creation of ______________ gave the world its largest free market (with a total output of about $6 trillion).

a. the European Union
b. the Association of Southeast Asian Nations (ASEAN)
c. the North American Free Trade Agreement (NAFTA)
d. the North Atlantic Treaty Organization (NATO)
e. the North and South American Group (NSAG)

Answer: (c) Difficulty: (3)

20. Some experts predict a global economy will evolve around a triad of three economic hubs. The three economic entities would be best described as being ____________________

a. North America, Asia, and Africa.
b. North America, the European Union, and Africa.
c. North America, Indian sub-basin, and China.
d. North and South America, the European Union, and Australia.
e. North America, the European Union, and East and Southeast Asia (including Japan, China, and India).

Answer: (e) Difficulty: (1)

21. Sources show that the European Union has gained strength in recent years. Many fear that the European Union will lead to a "fortress Europe to which non-EU firms will find it increasingly difficult to export goods." The solution seems to be to form ____________________

a. cartels.
b. joint ventures with European partners.
c. joint ventures with the Japanese.
d. joint ventures with Russia.
e. strategic alliances with non-European partners.

Answer: (b) Difficulty: (2)

22. All of the following are often cited as MAIN reasons that firms expand internationally EXCEPT ____________________

a. sales expansion.
b. finding "cheap labor."
c. seeking out foreign products and services to reduce their costs.
d. seeking high-quality goods and services.
e. smoothing out sales and profit swings.

Answer: (b) Difficulty: (3)

23. A manufacturer of snowblowers might sell in Chile knowing that, as demand for its products in the United States is dropping off in the spring, it is rising in Chile where the seasons are reversed. This would be an example of which of the following reasons that companies choose to go abroad?

a. sales expansion.
b. finding "cheap labor."
c. seeking out foreign products and services to reduce their costs
d. seeking high quality goods and services.
e. smoothing out sales and profit swings.

Answer: (e) Difficulty: (2)

24. Apple Computer recently enlisted the aid of Sony to produce parts for its new notebook computer. This would be an example of which of the following reasons that companies choose to go abroad?

a. sales expansion
b. finding "cheap labor"
c. seeking out foreign products and services to reduce their costs
d. seeking high-quality goods and services
e. smoothing out sales and profit swings

Answer: (d) Difficulty: (2)

25. Firms can use several strategies to expand internationally. All of the following are PRIMARY alternatives EXCEPT ______________________

a. exporting.
b. licensing.
c. franchising.
d. foreign direct investment.
e. expropriation.

Answer: (e) Difficulty: (2)

26. Which of the following international expansion strategies is often the first step when manufacturers decide to expand their sales abroad?

a. exporting
b. licensing
c. franchising
d. foreign direct investment
e. joint ventures

Answer: (a) Difficulty: (1)

27. ______________ means selling abroad, either directly to target customers or indirectly by retaining foreign sales agents and distributors.

a. Exporting
b. Licensing
c. Franchising
d. Foreign direct investment
e. Joint ventures

Answer: (a) Difficulty: (1)

28. Which of the following would be a PRIMARY advantage of exporting?

a. It protects patents and processes
b. It controls the use of products and services by foreign companies
c. It is a quick and "relatively inexpensive" way to go international.
d. Quality control can be maintained.
e. Pricing regulations and strategies can be maintained.

Answer: (c) Difficulty: (2)

29. ____________________ is an arrangement whereby a firm grants a foreign firm the right to use intangible property (such as patents, copyrights, manufacturing processes, or trade names) for a specified period of time (usually in return for a percentage of the earnings).

a. Exporting
b. Licensing
c. Franchising
d. Foreign direct investment
e. Joint ventures

Answer: (b) Difficulty: (1)

30. One of the chief DISADVANTAGES of using licensing as a means to expand a firm's business internationally is ____________________

a. the firm (by licensing its knowledge and know-how) can lose control over its intellectual property.
b. profit potential.
c. the firm's name is lost.
d. the firm can be accused of monopolistic practice.
e. the process may be too expensive to undertake.

Answer: (a) Difficulty: (1)

31. ________________ is the granting of a right by a parent company to another firm to do business in a prescribed manner.

a. Exporting
b. Licensing
c. Franchising
d. Foreign direct investment
e. Joint ventures

Answer: (c) Difficulty: (2)

32. All of the following are differences between licensing and franchising EXCEPT ________________

a. franchising involves a longer period of time.
b. franchising usually requires the adherence to stricter guidelines in running the business.
c. licensing is usually limited to manufacturers.
d. franchising is usually popular with service-oriented firms.
e. licensing involves a longer period of time.

Answer: (e) Difficulty: (2)

33. One of the significant DISADVANTAGES of franchising is that it is difficult for the franchise operator to ________________

a. make any money.
b. do advertising.
c. maintain quality control.
d. find and hire good managers.
e. protect the corporate logo or trademark.

Answer: (c) Difficulty: (3)

34. ____________________ refers to operations in one country that are controlled by entities in a foreign country. For example, the foreign firm might build new facilities in another country.

a. Exporting
b. Licensing
c. Franchising
d. Foreign direct investment
e. Joint ventures

Answer: (d) Difficulty: (1)

35. When Matsushita bought control of Rockefeller Center in New York City a few years ago, its actions became an example of which of the following expansion strategies listed below?

a. exporting
b. licensing
c. franchising
d. foreign direct investment
e. joint ventures

Answer: (d) Difficulty: (1)

36. A ________________ is referred to as a cooperative agreement between potential or actual competitors.

a. license agreement
b. franchise agreement
c. direct investment in a foreign environment
d. strategic alliance
e. joint venture

Answer: (d) Difficulty: (3)

37 An example of a ________________ occurred several years ago when aircraft manufacturing giant Boeing combined with a consortium of Japanese companies to produce the 767 commercial jet airliner.

a. license agreement
b. franchise agreement
c. direct investment in a foreign environment
d. strategic alliance
e. joint venture

Answer: (d) Difficulty: (2)

38. A ________________ is the participation of two or more companies in an enterprise in which each party contributes assets, owns the entity to some degree, and shares risk.

a. license agreement
b. franchise agreement
c. direct investment in a foreign environment
d. strategic alliance
e. joint venture

Answer: (e) Difficulty: (2)

39. An example of a ________________ occurred when General Motors joined Toyota and formed New United Motor Manufacturing, Inc. (NUMMI). Toyota got to sharpen its U.S. marketing skills and gain access to U.S. customers. General Motors was able to learn about dynamic Japanese manufacturing methods and technology.

a. license agreement
b. franchise agreement
c. direct investment in a foreign environment
d. strategic alliance
e. joint venture

Answer: (e) Difficulty: (2)

40. All of the following are advantages of a joint venture <u>EXCEPT</u> ________________

a. it lets partners gain experience in markets that may be foreign to one or more of the parties without assuming complete risk.
b. it lets a foreign partner combine talents with a local partner for combined strength.
c. costs can be shared.
d. industrial secrets can be shared without fear of infringement or theft.
e. through sharing, control risks become shared.

Answer: (d) Difficulty: (2)

41. In a _______________, ownership is 100 percent by the foreign firm.

a. license agreement
b. wholly owned subsidiary
c. direct investment in a foreign environment
d. strategic alliance
e. joint venture

Answer: (b) Difficulty: (1)

42. Which of the following foreign expansion forms provides for the tightest controls by the foreign firm?

a. license agreement
b. wholly owned subsidiary
c. direct investment in a foreign environment
d. strategic alliance
e. joint venture

Answer: (b) Difficulty: (2)

43. Which of the following foreign expansion forms is generally the least risky and least expensive for the firm wishing to expand into a foreign market. This form generally provides the least control over the market or future of its products in the foreign market.

a. exporting
b. wholly owned subsidiary
c. direct investment in a foreign environment
d. strategic alliance
e. joint venture

Answer: (a) Difficulty: (1)

44. The phrase "the tastes and preferences of consumers in different nations are beginning to converge on some global norm" exemplifies which of the following terms?

a. cultural differences
b. global marketing
c. market segmentation
d. product differentiation
e. product standardization

Answer: (b) Difficulty: (3)

45. All of the following are examples of honoring cultural differences EXCEPT ________________

a. making all cars with the steering wheel on the left side.
b. changing the spelling of key words or phrases to avoid negative meanings in foreign languages.
c. adding or deleting alcoholic beverages in certain markets dependent on customs.
d. adding or deleting beef products in certain markets dependent on religious customs.
e. changing the role of female models in advertisements in Muslim countries.

Answer: (a) Difficulty: (1)

46. ________________ means putting parts of a firm's production process in various locations around the globe.

a. Product standardization
b. Service standardization
c. Assembly line management
d. Globalizing production
e. Globalizing innovation

Answer: (d) Difficulty: (2)

47. In the late 1970s and early 1980s, Xerox Corporation's various companies were fragmented. In each country in which Xerox operated there was a separate structure of suppliers, assembly plants, and distribution channels. This would be an example of a firm that had yet to realize the benefits of ____________________

a. product standardization.
b. service standardization.
c. globalizing production.
d. assembly-line management.
e. globalizing innovation.

Answer: (c) Difficulty: (3)

48. When 3M makes tapes, chemicals, and electrical parts in Bangalore, India, and Hewlett-Packard assembles computers and designs memory boards in Guadalajara, Mexico, these would be examples of ____________________ in action.

a. production standardization
b. market-share dominance
c. global staffing
d. service dominance
e. competitive advantage

Answer: (c) Difficulty: (3)

49. Managers must carefully weigh behavioral factors when staffing their far-flung production facilities. All of the following would be important considerations (in the area of OB) in setting up a factory in Mexico EXCEPT ________________

a. workplace harmony.
b. role position.
c. status in the workplace.
d. religion of the employees.
e. the ability to exercise authority.

Answer: (d) Difficulty: (1)

50. A ________________ is a manager who views markets and production globally and who seeks higher profits for his or her firm on a global basis.

a. production-based manager
b. dominant manager
c. regional manager
d. cultural manager
e. global manager

Answer: (e) Difficulty: (1)

51. A good definitional phrase that would describe a "cosmopolitan" manager would be ____________

a well read.
b. innovative and spirited.
c. belonging to the world.
d. geographically aware.
e. biased against foreigners.

Answer: (c) Difficulty: (2)

52. A manager's philosophy about international business will influence his or her willingness to take a company goal. Several approaches are possible. Which of the following approaches would be most consistent with a "home-market-oriented firm or philosophy?"

a. ethnocentric
b. polycentric
c. regiocentric
d. geocentric
e. duocentric

Answer: (a) Difficulty: (2)

53. A manager's philosophy about international business will influence his or her willingness to take a company goal. Several approaches are possible. Which of the following approaches would be most consistent with a "company that is limited to several individual foreign markets?"

a. ethnocentric
b. polycentric
c. regiocentric
d. geocentric
e. duocentric

Answer: (b) Difficulty: (2)

54. Going international presents the manager with new and perplexing problems. All of the following are important areas in which the manager must develop skills and understanding EXCEPT ____________

a. economic.
b. legal.
c. sociocultural.
d. public relations.
e. technological.

Answer: (d) Difficulty: (3)

55. At the present time ______________ is an example of a pure-market economy. In a pure-market economy, the quantities and nature of the goods and services produced are not planned by anyone. Instead, the interactions of supply and demand in the market for goods and services determines what is produced, in what quantities, and at what prices.

a. the People's Republic of China
b. Hong Kong
c. Japan
d. United States
e. France

Answer: (b) Difficulty: (3)

56. Until very recently, ___________________ was an example of a command economy. In a command economy, government central planning agencies try to determine how much is to be produced by which sectors of the economy, and by which plants and for whom these goods are produced.

a. the People's Republic of China
b. Hong Kong
c. Japan
d. United States
e. France

Answer: (a) Difficulty: (2)

57. The exchange rate for one country's currency is the rate at which ______________

a. one country will make money.
b. arbitrage will occur.
c. it can be exchanged for another country's currency.
d. losses will occur.
e. central banking takes over.

Answer: (c) Difficulty: (1)

58. ______________ are (is) governmental influences usually aimed at reducing the competitiveness of imported products or services.

a. Export management
b. Import management
c. World trade
d. Trade controls
e. Embargoes

Answer: (d) Difficulty: (2)

59. The most common type of a trade control (governmental influences usually aimed at reducing the competitiveness of imported products or services) is a(n) ______________

a. embargo.
b. confiscation.
c. domestication.
d. expropriation.
e. tariff.

Answer: (e) Difficulty: (1)

60. There are many important differences between legal systems in countries in the global environment. One of these important differences concerns the type of law enforced. ______________ is based on tradition and depends more on precedence and customs than on written statutes.

a. Code law
b. Common law
c. Royal law
d. Declared law
e. International law

Answer: (b) Difficulty: (2)

61. An example of a country that depends heavily the use of common law (the system that depends on tradition, precedence, and customs rather than written statutes) is ____________________

a. England.
b. France.
c. Algeria.
d. Morocco.
e. Quebec province of Canada.

Answer: (a) Difficulty: (3)

62. People who travel to other countries quickly learn that they must also adapt to societal differences. In ______________, for example, indirect and vague communication is preferred, with sentences frequently left unfinished so the other person may draw his or her own conclusions.

a. Germany
b. France
c. India
d. United States
e. Japan

Answer: (e) Difficulty: (2)

63. Research by Geert Hofstede suggests that a society's values are among the most influential cultural differences. Hofstede argues that cultural differences among countries reflect all of the following basic values EXCEPT ______________

a. language uncertainty.
b. power distance.
c. individualism versus collectivism.
d. masculinity versus femininity.
e. uncertainty avoidance.

Answer: (a) Difficulty: (2)

64. Research by Geert Hofstede suggests that a society's values are among the most influential cultural differences. According to Hofstede, the ________________ value would be of importance if "all members of a country were expected to look after themselves and their immediate families."

a. language uncertainty
b. power distance
c. individualism versus collectivism
d. masculinity versus femininity
e. uncertainty avoidance

Answer: (c) Difficulty: (2)

65. Research by Geert Hofstede suggests that a society's values are among the most influential cultural differences. According to Hofstede, the ________________ value would be of importance if "people in a society are uncomfortable with unstructured situations in which unknown, surprising, or novel incidents occur."

a. language uncertainty
b. power distance
c. individualism versus collectivism
d. masculinity versus femininity
e. uncertainty avoidance

Answer: (c) Difficulty: (2)

66. A country's technological environment can determine a product's success abroad. The ease or difficulty with which technology can be transferred from one country to another is of constant concern. Successful technology transfer depends on all of the following EXCEPT ______________

a. a suitable technology to be transferred.
b. social and economic conditions must favor the transfer.
c. a willingness to accept and adapt the technology.
d. a patent office system so technology can be protected.
e. an ability to accept and adapt the technology.

Answer: (d) Difficulty: (3)

67. ______________ means setting goals and identifying the courses of action for achieving those goals.

a. Organizing
b. Planning
c. Control
d. Coordinating
e. Leading

Answer: (b) Difficulty: (1)

68. Companies evolve from domestic to multinational enterprises in countries around the world. During this process, several organizational stages occur. With respect to these alternatives, if a division handles its own foreign sales (which may come largely from unsolicited overseas orders) it is called a(n) ________________

a. domestic or traditional division organization.
b. export-oriented division structure.
c. international division structure.
d. multinational division structure.
e. alliance structure.

Answer: (a) Difficulty: (2)

69. Companies evolve from domestic to multinational enterprises in countries around the world. During this process, several organizational stages occur. With respect to these alternatives, if a division coordinates all international activities such as licensing, contracting, and managing foreign sales, it is called a(n) ______________

a. domestic or traditional division organization.
b. export-oriented division structure.
c. international division structure.
d. multinational division structure.
e. alliance structure.

Answer: (b) Difficulty: (2)

70. Companies evolve from domestic to multinational enterprises in countries around the world. During this process, several organizational stages occur. With respect to these alternatives, if a division is divided into separate domestic and international divisions (where the international division focuses on production and sales overseas, while the domestic division focuses on domestic markets), it is called a(n) ______________

a. domestic or traditional division organization.
b. export-oriented division structure.
c. international division structure.
d. multinational division structure.
e. alliance structure.

Answer: (c) Difficulty: (2)

71. _____ means monitoring actual performance to ensure it is consistent with the standards expected.

a. Organizing
b. Planning
c. Control
d. Coordinating
e. Leading

Answer: (c) Difficulty: (1)

<u>True/False and Short Answer</u>

72. The terms international business, international trade, and international investment have all come to mean the same thing in the contemporary management environment of the late 20th century.

Answer: (False) Difficulty: (2)

73. International business refers to all business transactions that involve two or more countries.

Answer: (True) Difficulty: (1)

74. ____________________ is the performance of the management functions of planning, leading, and controlling across national borders.

Answer: (International management) Difficulty: (3)

75. A __________________ may be defined as being an internationally integrated business that is controlled by a parent corporation, owned and managed essentially by the nationals of its home country.

Answer: (multinational corporation [MNC]) Difficulty: (3)

76. General Electric, General Motors, and ITT would be examples of multinational corporations.

Answer: (True) Difficulty: (1)

77. The multinational, transnational, or global corporation are (for all intents and purposes) the same thing.

Answer: (False) Difficulty: (2)

78. _____________ means all trade barriers among participating countries are removed.

Answer: (Free trade) Difficulty: (2)

79. The classic explanation for the advantages of free trade was presented in The Wealth of Nations written by John Maynard Keynes.

Answer: (False) Difficulty: (1)

80. In a __________________, all barriers to trade among member countries are removed.

Answer: (free-trade area) Difficulty: (2)

81. In a common market, all members are free to trade with one another, may collect tariffs or duties at their borders from members, and band together to prohibit unfair trade from outside the market.

Answer: (False) Difficulty: (2)

82. The world's largest free market is NAFTA.

Answer: (True) Difficulty: (1)

83. Experts predict that in the next century three global economic hubs will control and share the world's economic resources. The hubs are projected to be North America, the Sino-Soviet group, and Southeast Asia.

Answer: (False) Difficulty: (1)

84. Firms expand internationally for several reasons. Finding cheap labor is usually not a main motive for most firms internationalization efforts.

Answer: (True) Difficulty: (2)

85. Firms may choose from several strategies when they wish to expand internationally. Often, the first strategy chosen in franchising.

Answer: (False) Difficulty: (2)

86. One of the easiest and least risky ways to test the waters in the host country and learn more about customers' needs is to export.

Answer: (True) Difficulty: (1)

87. ________________ refers to operations in one country that are controlled by entities in a foreign country.

Answer: (Foreign direct investment) Difficulty: (3)

88. A________________ is the participation of two or more companies in an enterprise in which each party contributes assets, owns the entity to some degree, and shares risk.

Answer: (joint venture) Difficulty: (3)

89. Global managers today must be cosmopolitan.

Answer: (True) Difficulty: (i)

90. An ethnocentric management philosophy may manifest itself as one where the company limits itself to several individual foreign markets as a primary strategy.

Answer: (False) Difficulty: (2)

91. Research by Geert Hofstede suggests that a society's values are among the most influential cultural differences. One of his important "values" measurements is masculinity versus femininity. According to his studies, Japan and Austria ranked high on the masculinity scale.

Answer: (True) Difficulty: (2)

Essay

92. Demonstrate the differences between the following terms: international business, international trade, international investment, international management, the multinational corporation, and the global or transnational corporation. You may define it you so choose, however, be sure to note the differences between the terms.

Answer:

a. International business: any firm that engages in international trade or investment.

b International trade: the export or import of goods or services to consumers in another country.

c. International investment: the investment of resources in business activities outside a firm's home country.

d. International management: the performance of the management functions of planning, organizing, leading, and controlling across national borders.

e. Multinational corporation (MNC): an internationally integrated business that is controlled by a parent corporation, owned and managed essentially by the nationals of its home country.

f. Global or transnational corporation: according to Theodore Levitt, the multinational corporation is being replaced by the global or transnational corporation. This form operates as if the entire world (or major regions of it) were a single entity and sells essentially the same things in the same way everywhere.

Difficulty: (2)

93. The concept of free trade helps to explain why business today is increasingly international. In a short discussion, explain what free trade and economic integration are. In addition, discuss the concepts of a free-trade area, a customs union, and a common market. Give a short illustration or example of each of the three concepts.

Answer:

Free trade means that all trade barriers among participating countries are removed. The classic illustration for this concept came in Adam Smith's The Wealth of Nations. He argued that each country would, if unhindered by subsidies and tariffs, end up specializing in the goods and services that it could produce best.

Economic integration occurs when two or more nations obtain the advantages of free trade by minimizing trade restrictions between them. Economic integration occurs on several levels. First, in a free-trade area, all barriers to trade among member countries are removed, so that goods and services are freely traded among the member countries. Second, at the next highest level of integration, members of a customs union dismantle barriers to trade among themselves while establishing a common trade policy with respect to non-members. Lastly, in a common market, no barriers to trade exist among members and a common external trade policy is in force; in addition, factors of production, such as labor, capital, and technology, move freely between member countries.

Examples would be the European Union, NAFTA, and ASEAN (Southeast Asian Nations organization).

Difficulty: (2)

94. Briefly identify and discuss each of the primary types of international expansion strategies.

Answer:

a. Exporting: often the first strategy used. Exporting means selling abroad, either directly to target customers or indirectly by retaining foreign sales agents and distributors. More than one-half of all world trade is done in this manner. This is a relatively quick and inexpensive way to go international and it is a good way to test the international waters and learn more about international consumers. On the other hand, costs (such as in transportation, tariffs, or manufacturing) can put the exporter at a disadvantage (especially if intermediaries are not chosen wisely).

b. Licensing: This is an arrangement whereby a firm (the licensor) grants a foreign firm the right to use intangible property (such as patents, copyrights, manufacturing processes, or trade names) for a specified period of time in return for a percentage of the earnings (a royalty). The advantage is that one can enter a market for relatively low cost and receive exposure. However, lack of control and the possibility of losing trade secrets and know-how can outweigh the advantages of licensing.

c. Franchising: granting the right by a parent company to another firm to do business in a prescribed manner. This is similar to licensing but usually involves a longer period of time. In addition, stricter guidelines are usually imposed on the franchisee. The advantages and disadvantages are basically the same as those of licensing, however, maintaining quality and image control is a significant problem (such as at a McDonald's).

d. Foreign Direct Investment: refers to operations in one country that are controlled by entities in a foreign country (such as building new manufacturing facilities in another country). Or a foreign firm might acquire control of a domestic firm and accomplish the same purpose. Risk rises with this form of expansion.

e. Joint ventures (strategic alliances): Strategic alliances refer to cooperative agreements between potential or actual competitors, whereas joint ventures are the joint participation of two or more companies in an enterprise in which each party contributes assets, owns the entity to some degree, and shares risk. Both of these forms enable the participants to share risk, profits, and goals. However, like all partnerships, unequal relationships can cause problems. In addition, many of the partners in these ventures are suspicious of one another because they are competitors in other markets or industries.

f. Wholly Owned Subsidiaries: As the name implies, 100 percent ownership is by the foreign firm. These forms provide the greatest control, but they may be very costly ways to enter a market. In addition, all assets can be lost in a hostile environment.

Difficulty: (3)

95. Research by Geert Hofstede suggests that a society's values are among the most influential cultural differences. Values are basic beliefs we hold about what is good or bad, important or unimportant. Hofstede suggests four basic values that are involved in the cultural differences between countries and international markets. Identify each of these areas and briefly summarize the primary characteristics of each value area. Be specific in your comments.

Answer:

a. Power distance: refers to the extent to which the less powerful members of institutions accept and expect that power will be distributed unequally. Hofstede concluded that the institutionalization of such an inequality was higher in some countries (such as Mexico) than in others (such as Sweden).

b. Individualism versus collectivism: The degree to which ties between individuals are normally loose or close is measured as individualism or collectivism. In more individualistic countries, "all members are expected to look after themselves and their immediate families" (ex. United States and Australia). In collectivist countries, people are expected to care for each other more (ex. Indonesia).

c. Masculinity versus femininity: societies differ in the extent to which they value assertiveness (masculinity) or caring (femininity). Masculine countries are exemplified by Japan and Austria. Feminine countries are exemplified by Denmark and Costa Rica.

d. Uncertainty avoidance: This concept refers to whether people in the society are uncomfortable with unstructured situations in which unknown, surprising, or novel incidents occur. How do consumers deal with surprises? High on the scale for dealing with these events are people in Sweden, Israel, and Great Britain. Low on the scale are people living in Greece and Portugal.

Difficulty: (2)

96. As companies evolve from domestic to multinational enterprises, they typically go through several organizational stages. Their increasing international operations often necessitate a more globally oriented organizational design. Take each of the four forms mentioned in your chapter and characterize each. Be careful to point out differences between the forms if the differences are not revealed in definitions.

Answer:

a. Traditional division structure: Here, each division handles the minimal existing sales to each market itself. These sales may largely come from unsolicited overseas orders.

b. Export-oriented division structure: As a response to increasing foreign orders, the firm re-organizes into this format. Here the export division or department helps to coordinate foreign sales, for instance by assisting with freight-forwarding arrangements.

c. International division structure: Here, each domestic division produces and sells for the local market. The international division may handle production and sales for foreign markets, or possibly just market-abroad products. In general, the domestic division takes care of domestic issues and the foreign division focuses on foreign opportunities and situations.

d. Multinational division structure: Here, each country of a geographic region is the responsibility of its own officer and division, which handles sales and production within its own geographic boundaries. Increasingly, companies are consolidating production facilities into more efficient units.

Difficulty: (3)

Chapter 3

MANAGING ETHICS, DIVERSITY, AND SOCIAL RESPONSIBILITY

INTRODUCTION

The management of ethics, diversity, and social responsibility is important for organizations to be successful in the complex work place, today. This chapter explains the nature of ethical decision making, the factors that influence the making of ethical or unethical decisions, and examples of ethics programs in organizations. Approaches to social responsibility are analyzed. Techniques for managing workforce diversity are discussed to provide greater understanding of people in organizations, today.

LECTURE OUTLINE

What's Ahead?

The Ethical Challenge of Work
- The Nature of Ethics
- Ethics and the Law

What Determines Whether Behavior Is Ethical at Work?
- The Individual Applies His or Her Own Standards
- The Organization Shapes Individuals' Behavior
- The Leader Influences Others' Ethics
- Ethics Policies and Codes Make Standards Concrete

Ethics Programs in Practice
- The Johnson & Johnson Program
- The Norton Company
- In Summary: How to Foster Ethics at Work

Social Responsibility at Work
- Ethics and Social Responsibility
- To Whom Should the Corporation Be Responsible?
- How to Improve Social Responsiveness

Managing Diversity
- Bases for Diversity
- Barriers in Dealing with Diversity
- Boosting Performance by Managing Diversity

Summary

Case: Apex Fuel Oil's Battle Against Drugs
You Be the Consultant: Ethics and ConEdison

ANNOTATED OUTLINE

What's Ahead?

Volkswagen won a major coup in the automotive wars when they hired GM's purchasing and production whiz, Jose Ignacio Lopez. It soon became apparent that Lopez took a number of key assistants with him. The ethical situation became clouded, however, when GM alleged that Lopez also took many important GM plans, contracts, and other sensitive documents. How should VW respond to these accusations, and if they prove true, what should be done about Senior Lopez?

Notes: How did this case conclude?

The Ethical Challenge at Work

The Nature of Ethics. Ethics refers to the principles of conduct governing an individual or a group and specifically to the standards used to govern conduct. Ethics also includes morality or accepted norms of behavior in society.

Notes:

Ethics and the Law. The law is not an adequate guide to whether a decision is ethical. Decisions can be legal but not ethical. They can also be ethical and illegal.

Notes:

What Determines Whether Behavior Is Ethical at Work?

The Individual Applies His or Her Own Standards. Individuals are accountable for ethical or unethical decisions in the work place because they bring their own ideas of what is morally right and wrong. Studies have shown that older workers have stricter interpretations of ethical standards and make more ethical decisions than younger employees. Perhaps lack of experience of younger employees makes them more open to making the wrong decision.

Notes:

The Organization Shapes Individuals' Behavior. Unethical business practice involves the tacit, if not explicit, cooperation of others and reflects the values, attitudes and beliefs, language and behavioral patterns that define the culture of an organization. Rarely is corporate misconduct the result of only one employee's character flaws. The Sears Auto Center case is an example of how good intentions can create conditions in which unethical decisions can flourish.

Notes:

The Leader Influences Others' Ethics. Ethical decisions can be influenced knowingly and unknowingly by supervisors who need to be careful regarding the signals their actions or inactions send to subordinates. Leaders should clarify expectations for employees, be careful about the use of signs and signals, and use rites or ceremonies to reinforce desired behavior.

Notes:

Ethical Policies and Codes Make Standards Concrete. Organizations need an ethics policy to foster corporate behavior of a high standard. The ethics policy signals that top management is serious about ethics and wants to foster a culture that takes ethics seriously. One study found that corporate ethical codes of conduct were correlated with the size of the organization, with larger firms more likely to have codes in place. Fines imposed upon corporations for unethical conduct are generally lower if the firm had an ethics policy and could show it made an effort to ensure ethical behavior.

Notes:

Ethics Programs in Practice

The Johnson & Johnson Program. The company demonstrated that its credo is more than just words when it recalled Tylenol during the poison scare and lost hundreds of millions of dollars in sales. The credo has been described as the "glue" that holds the firm together by providing a common focus and set of standards for its operations globally.

Notes:

The Norton Company. The Norton Ethics Program has a code of conduct, an ethics committee of the board of directors, and an annual ethics review. The code contains general guidelines regarding responsibility to employees, customers, suppliers, shareholders, local communities, and the general public. Compliance is monitored through subordinate performance reviews and upward reporting from the top 100 managers each year.

Notes:

In Summary: How to Foster Ethics at Work. One study concluded that fostering ethics at work involved emphasizing top-management commitment, publishing a code, establishing compliance mechanisms, involving personnel at all levels, and measuring results. Surveys or audits should be used to monitor compliance with ethical standards.

Notes:

Social Responsibility at Work

Ethics and Social Responsibility. Corporate social responsibility refers to the extent to which companies should and do channel resources toward improving one or more segments of society other than the firm's own stockholders. Social responsibility is an ethical issue because it involves questions of what is morally right or wrong regarding the responsibilities of the firm.

Notes:

To Whom Should the Corporation Be Responsible? Managerial capitalism is the classical view that a corporation's primary purpose is to maximize profits for its stockholders. An opposite view is the stakeholder theory that business has a social responsibility to serve all the corporate stakeholders affected by its business decisions. The moral minimum is in between managerial capitalism and stakeholder theory. It holds that the purpose of the corporation is to maximize profits, but that in so doing, it will commit no harm. Ben & Jerry's Homemade is an example of a firm that links social responsibility with the purchase of ingredients for its products.

Notes:

How to Improve Social Responsiveness. Corporate social audits and whistle blowing can be used to help a firm improve its social responsiveness by providing communication that will enable management to take action as soon as possible when wrongdoing or potential wrongdoing is exposed.

Notes:

Managing Diversity

Bases for Diversity. Managing diversity means planning and implementing organizational systems and practices to manage people so that the potential advantages of diversity are maximized while its potential disadvantages are minimized. A workforce is diverse when it is comprised of two or more groups, each of whose members are identifiable and distinguishable based on demographic or other characteristics. Diversity is usually based on racial and ethnic groups, women, older workers, and people with disabilities or varying sexual/affectional orientation.

Notes:

Barriers in Dealing with Diversity. Important barriers that may prevent a company from taking full advantage of the potential in its diverse workforce include stereotyping and prejudices, ethnocentrism, discrimination, tokenism, and gender-role stereotypes. Stereotyping involves ascribing behavioral traits to individuals based on their apparent membership in a group, while prejudice is a bias resulting from prejudging someone on the basis of some trait. Ethnocentrism is a tendency to view members of one's own group as the center of the universe and viewing other social groups less favorably than one's own. Discrimination refers to a behavioral bias toward or against the person based on the person's group. Tokenism is the tendency of some companies to appoint a relatively small group of women or minority-group members to high-profile positions rather than more aggressively achieving full work-group representation for that group. Gender-role stereotyping is the tendency to associate women with certain jobs.

Notes:

Boosting Performance by Managing Diversity. Diversity management involves compulsory and voluntary management actions. Many legally mandated actions are required to minimize discrimination at work. Additionally, strong leadership must include taking a strong personal stand on the need for change while providing financial and other support need to implement actual changes. Research should be conducted to provide information about employees' current attitudes and perceptions toward different groups within the company. Diversity training and education should be provided on a regular basis.

Educational programs and other approaches should be implemented to change the organization's culture and management systems to foster more effective management of diversity. The managing diversity efforts must be monitored and evaluated continuously with changes implemented where necessary to maintain its effectiveness.

Notes:

Summary

Notes:

Case: Apex Fuel Oil's Battle Against Drugs

1. The issue of employee right to privacy clashes with the obligation of the organization to provide a safe work place. The company has handled the issues in a humane way. It offers employees the opportunity for rehabilitation while still receiving a substantial amount of salary that decreases as the length of time increases during rehabilitation. The program apparently has been successful as evidenced by lower worker's compensation costs and insurance rates. These cost savings help offset the costs of rehabilitation.

2. The partners had to consider the interests of employees and their families, citizens traveling the highways, the shareholders, and the interests of the government. Managers, customers, and suppliers also have a stake in the decision because, if the problems remained unchecked, the viability of the firm could have been threatened.

3. The company has social responsibility to its employees to provide a safe work place and to the community by ensuring that its drivers are drug-free and not a hazard on the highways. It also has social responsibility to its shareholders to manage the operation in their interests. The partners did the socially responsible thing by attempting to rehabilitate drug abusers rather than immediately discharging them.

You Be the Consultant: Ethics and ConEdison

1. The decision to cut off the power may have been ethical, but the manner in which it was implemented was completely unethical. Con Edison should have told the firm it would turn off its power on the day in question if money due had not been remitted. Its representative should have told management why he was there and given the firm one last opportunity to pay before turning off the power.

2. Dorf could have denied access to the property to the Con Edison employee until the following Monday. It certainly was odd that an electric utility employee would appear at a customer's location at 5:45 p.m. on a Friday afternoon to conduct business. With the Knitting Factory preparing for a busy weekend, it would not have been unusual or unethical for Dorf to ask the man to return on Monday.

End of Chapter Questions for Chapter 3

Multiple Choice

1. Sales executives at an artificial eye-lens firm offered doctors "free use of a yacht off Florida, travel in Europe, all-expenses-paid, week-long training seminars in the Bahamas, second homes, and cash rebates for buying their lenses rather than their competitors'." This would be an example of a(n) ________________ facing contemporary managers.

a. tax rebate scheme
b. attempt to stimulate demand in the public sector
c. ethical choice
d. legitimate promotional process
e. strategic leadership choice

Answer: (c) Difficulty: (2)

2. Senior General Dynamics managers allegedly added $64 million of improper overhead expenses (including country-club memberships and dog kennel fees) to their company's bill for defense contracts in the early 1980s. This would be an example of a(n) ________________ facing contemporary managers.

a. tax rebate scheme
b. attempt to stimulate demand in the public sector
c. ethical choice
d. legitimate promotional process
e. strategic leadership choice

Answer: (c) Difficulty: (1)

3. ____________ refer(s) to the principles of conduct governing an individual or a group and (specifically) to the standards used to govern conduct.

a. Judgment
b. Conduct
c. Principles
d. Morals
e. Ethics

Answer: (e) Difficulty: (2)

4. Ethics always involve ______________ and ______________ judgments. Select the most complete answer offered below.

a. right and wrong
b. normative and non-normative
c. monetary and non-monetary
d. religious and non-religious
e. simple and complex

Answer: (b) Difficulty: (2)

5. A _______________ judgment implies "that something is good or bad, right or wrong, or better or worse."

a. non-normative
b. simple
c. complex
d. normative
e. reference

Answer: (d) Difficulty: (3)

6. That's a great outfit" is a statement that would apply to a _______________ judgment. Select the most complete phrase.

a. non-normative
b. simple
c. complex
d. normative
e. reference

Answer: (d) Difficulty: (3)

7. "You are wearing a skirt and blouse" is a statement that would apply or call for a _______________ judgment. Select the most complete phrase.

a. non-normative
b. simple
c. complex
d. normative
e. reference

Answer: (a) Difficulty: (3)

8. Ethical decisions almost always involve _______________, in other words, society's accepted norms of behavior.

a. judgments
b. references
c. morality
d. religion
e. standards

Answer: (c) Difficulty: (2)

9. Moral standards differ from non-moral ones in all of the following ways EXCEPT ____________

a. moral standards are always open to reference and inference.
b. moral standards address matters of serious consequence to human well-being.
c. moral standards cannot be established or changed by decisions of authoritative bodies.
d. moral standards should override self-interest.
e. moral judgments are never situational.

Answer: (a) Difficulty: (3)

10. If you have a moral obligation to do something and you are expected to do it even if doing so might (for instance) impede your career or cost you money, this is an example of usage of moral standards or judgments from which of the following areas?

a. Moral standards tend to trigger strong emotions.
b. Moral standards address matters of serious consequence to human well-being.
c. Moral standards cannot be established or changed by decisions of authoritative bodies.
d. Moral standards should override self-interest.
e. Moral judgments are never situational.

Answer: (d) Difficulty: (2)

11. The fact that your company will benefit if you pay a bribe does not make the bribe right. The preceding statement is an example of usage of moral standards or judgments from which of the following areas?

a. Moral standards tend to trigger strong emotions.
b. Moral standards address matters of serious consequence to human well-being.
c. Moral standards cannot be established or changed by decisions of authoritative bodies.
d. Moral standards should override self-interest.
e. Moral judgments are never situational.

Answer: (e) Difficulty: (2)

12. Pick the best response to the following situation. A 38-year-old employee is fired just before he has earned the right to a pension.

a. This is probably illegal.
b. This is probably legal.
c. This is probably legal but unethical.
d. This is probably illegal and ethical.
e. This is probably illegal and unethical.

Answer: (c) Difficulty: (3)

13. Pick the best response to the following situation. A naive customer (obviously unfamiliar with the store or the merchandise) is charged an exorbitant price for merchandise.

a. This is probably illegal.
b. This is probably legal.
c. This is probably legal but unethical.
d. This is probably illegal and ethical.
e. This is probably illegal and unethical.

Answer: (c) Difficulty: (3)

14. Pick the best response to the following situation. A store uses infrared sensors in store ceilings to track shopping carts to determine customer's buying habits. A monitoring booth in the store records and maintains records of these data events.

a. This is probably illegal.
b. This is probably legal.
c. This is probably legal but unethical.
d. This is probably illegal and ethical.
e. This is probably illegal and unethical.

Answer: (c) Difficulty: (3)

15. Several factors influence whether specific people in specific organizations make ethical or unethical decisions. All of the following would be considered among those factors EXCEPT ______________

a. the individual applies his or her own standards.
b. the individual applies governmental standards.
c. the organization shapes individuals' behavior.
d. the leader influences others' ethics.
e. the organization develops ethics policies and codes.

Answer: (b) Difficulty: (2)

16. The ________________ should shoulder most of the credit (or blame) for any ethical decisions made by the employee (individual).

a. individual
b. leader
c. organization
d. department manager
e. supervisor

Answer: (a) Difficulty: (1)

17. Based on research findings, it has been concluded that ethical intentions (when given ethical problems to solve or respond to) of surveyed CEOs are more strongly affected by ________________ than by environmental pressures or organizational/situational characteristics.

a. their board of directors
b. their subordinates
c. their peers
d. their personal predispositions
e. the corporation's policies

Answer: (d) Difficulty: (3)

18. In a recent survey, 421 employees were surveyed to measure the degree to which certain characteristics influenced responses to ethical decision scenarios. In general, only ____________ proved to be a good predictor of whether a person would make the "right" decision.

a. age
b. gender
c. marital status
d. education
e. years in business

Answer: (a) Difficulty: (2)

19. Pick the best available completion to the following sentence about ethics and ethical conduct. Unethical business practice is most often the result of ________________

a. an individual's misconduct.
b. a supervisor's misconduct.
c. a CEO's misconduct.
d. a misunderstanding on the part of the employee or the supervisor.
e. several employees (possibly at varying levels in the organization) tacitly (if not explicitly) cooperating with others.

Answer: (e) Difficulty: (2)

20. In 1992, consumers and attorneys general in more than 40 states accused Sears of misleading customers; the specific complaint was that Sears service writers had sold customers unnecessary parts and services (from brake jobs to front-end alignments). These unethical business practices were probably the result of ________________

a. an individual's misconduct.
b. a supervisor's misconduct.
c. a CEO's misconduct.
d. a misunderstanding on the part of the employee or the supervisor.
e. several employees (possibly at varying levels in the organization) tacitly (if not explicitly) cooperating with others.

Answer: (e) Difficulty: (1)

21. In 1992, consumers and attorneys general in more than 40 states accused Sears of misleading customers; the specific complaint was that Sears service writers had sold customers unnecessary parts and services (from brake jobs to front-end alignments). The end result to the cases was interesting and showed the depth of difficulty that Sears had with its ethical practice. According to the best interpretation of the case, the end result was to conclude that ____________________

a. Sears was a crooked organization.
b. Sears was a misunderstood organization.
c. honest people with good intentions can create conditions in which unethical decisions can flourish.
d. competitive pressure justifies the means to the end.
e. all service businesses are just "a little unethical."

Answer: (c) Difficulty: (1)

22. It's usually not just what the leader says but what he or she does that shapes employees' decisions. The leader creates a particular ________________ and the employees then take their signals from it.

a. organizational psychology
b. organizational development process
c. team group
d. matrix structure
e. organizational culture

Answer: (e) Difficulty: (2)

23. A(n) ________________ is a basic belief about what you should and should not do, and what is and is not important.

a. attitude
b value
c. trait
d. ethical symbol
e. code

Answer: (b) Difficulty: (3)

24. Organizational culture is an important influence because a firm's culture reflects its ________________, and these ______________ help guide and channel people's behavior. (Choose one word of phrase to fill in both blanks.)

a. attitudes
b. feelings
c. codes
d. shared values
e. premonitions

Answer: (d) Difficulty: (1)

25. Leaders play a major role in creating and sustaining a firm's culture through the actions they take, the comments they make, and the visions they espouse. All of the following are specific ways in which leaders can shape their organization's culture EXCEPT ____________________

a. clarifying your expectations.
b. not expecting too much ethical behavior, but dealing with unethical behavior promptly.
c. using signs as guidelines.
d. using symbols as guidelines.
e. using rites and ceremonies.

Answer: (b) Difficulty: (2)

26. According to research, the single most important factor in fostering corporate behavior of a high ethical standard is ____________________

a. the actions of the employees.
b. the actions of supervisors.
c. the actions of the leader.
d. the actions of regulatory bodies.
e. the statement about ethics found in the corporate report.

Answer: (c) Difficulty: (3)

27. An ________________ signals that top management is serious about ethics and wants to foster a culture that takes ethics seriously.

a. ethics manager
b. ethics penalty
c. ethics book
d. ethics policy
e. ethics seminar

Answer: (d) Difficulty: (2)

28. Many firms have ethics codes. Research has shown that about 75 percent of all firms with a net worth greater than $100 million had corporate codes of conduct. Conclusions from research conducted with these firms indicated that all of the following were found to be true EXCEPT ____________________

a. top management must make it clear that it is serious about code enforcement.
b. ethics code violations must be dealt with quickly and publicly exposed.
c. it is important that the code of ethics be endorsed by executives at or near the top of the organization.
d. it is important that the code of ethics be endorsed by the employees for it to really be effective.
e. to influence employee behavior, the ethics code must be communicated.

Answer: (b) Difficulty: (2)

29. After a review of the ethics programs at eleven major firms, one study concluded that fostering ethics at work involved all of the following steps EXCEPT ________________

a. emphasizing the profit consequences of any ethical action to be taken.
b. emphasizing top management's commitment.
c. publishing a "code."
d. establishing compliance mechanisms.
e. involving personnel at all levels.

Answer: (a) Difficulty: (2)

30. After a review of the ethics programs at eleven major firms, one study concluded that fostering ethics at work involved five main steps. The first of these steps is best described as being to _______

a. emphasize the profit consequences of any ethical action to be taken.
b. emphasize top management's commitment.
c. publish a "code."
d. establish compliance mechanisms.
e. involve personnel at all levels.

Answer: (b) Difficulty: (3)

31. After a review of the ethics programs at eleven major firms, one study concluded that fostering ethics at work involved five main steps. In the experience of the firms surveyed with regard to corporate ethics, no point emerges more clearly than the crucial role of the firm's willingness to ________________

a. measure results.
b. emphasize top management's commitment.
c. publish a "code."
d. establish compliance mechanisms.
e. involve personnel at all levels.

Answer: (b) Difficulty: (3)

32. Firms with effective ethics programs set forth principles of conduct in a(n) __________. Examples of areas that might be included are fundamental honesty and adherence to laws, product safety and quality, health and safety in the work place, and conflicts of interest.

a. rule book
b. operator's manual.
c. legal manual.
d. strategic time line.
e. written code.

Answer: (e) Difficulty: (1)

33. Paying attention to values and ethics in recruiting and hiring, emphasizing corporate ethics in training, institution of communications programs to inform and motivate employees, and auditing ethical issues would all be examples of ______________ established by the organization to insure that ethical standards were being observed.

a. a rule book
b. an operator's manual
c. compliance mechanisms
d. strategic time lines
e. a written code

Answer: (c) Difficulty: (2)

34. If a firm were to use roundtable discussions among small groups of employees regarding corporate ethics and employees' surveys of attitudes regarding the state of ethics in the firm, the firm would be addressing which of the following steps for fostering ethics in the organization?

a. measure results
b. emphasize top management's commitment
c. publish a "code"
d. establish compliance mechanisms
e. involve personnel at all levels

Answer: (e) Difficulty: (3)

35. The normal method of measuring results of ethics programs in achieving compliance with ethical standards is to use ______________

a. tests.
b. observations.
c. case studies.
d. surveys or audits.
e. consultant-run focus groups.

Answer: (d) Difficulty: (1)

36. Corporate ______________ refers to the extent to which companies should and do channel resources toward improving one or more segments of society other than the firm's own stockholders.

a. diversity
b. ethics
c. minority programs
d. social responsibility
e. social globalization

Answer: (d) Difficulty: (2)

37. An example of socially responsible behavior would be to ____________________

a. have a company picnic to reward a great sales year.
b. create a program to reward employees for new product ideas.
c. create a program to reward employees for high attendance at the work place.
d. create a mentoring program.
e. creating jobs for minorities.

Answer: (e) Difficulty: (2)

38. Social responsibility is essentially a(n) ________________, since it involves questions of what is morally right or wrong with regard to the firm's responsibilities.

a. ethical issue
b. profitability issue
c. human rights issue
d. corporate rights issue
e. global issue

Answer: (a) Difficulty: (1)

39. Traditional ethical issues are often cited as being those of bribery, stealing, and corporate dishonesty. However, many perfectly ethical people in business strongly believe that a company's only social responsibility is to ______________________

a. its domestic consumers.
b. its foreign consumers.
c. its management.
d. its employees.
e. its shareholders.

Answer: (e) Difficulty: (3)

40. The classical view is that the corporation's primary (and perhaps sole) purpose is to ____________

a. be a good corporate citizen.
b. maximize profits for its stockholders.
c. create a productive and innovative work place for employees.
d. provide an environment that breeds security for senior executives.
e. minimize losses and maximize productivity.

Answer: (b) Difficulty: (2)

41. The classical view is that the corporation's primary (and perhaps sole) purpose is to maximize profits for its stockholders. Today, this view is most often associated with economist ____________.

a. John Stuart Mill.
b. John Maynard Keynes.
c. Milton Friedman.
d. Adam Smith.
e. Max Weber.

Answer: (c) Difficulty: (3)

42. The classical view is that the corporation's primary (and perhaps sole) purpose is to maximize profits for its stockholders. Consider the following statement: "Few trends could so thoroughly undermine the very foundation of our free society as the acceptance by corporate officials of a social responsibility other than to make as much money for their stockholders as possible." This view is most often associated with economist ______________.

a. John Stuart Mill.
b. John Maynard Keynes.
c. Milton Friedman.
d. Adam Smith.
e. Max Weber.

Answer: (c) Difficulty: (3)

43. The classical view is that the corporation's primary (and perhaps sole) purpose is to maximize profits for its stockholders. This view is built on two main arguments. Which of the following statements best describes one of those arguments?

a. Stockholders owe taxes; therefore, social responsibility has been paid.
b. It is the government's responsibility in a free-market economy to provide for social responsibility.
c. It is each individual's responsibility to provide for social responsibility.
d. Stockholders are owners of the corporation and so the corporate profits belong to them and to them alone.
e. If someone does not contribute, they should not receive rewards from society.

Answer: (d) Difficulty: (3)

44. The opposite view of the classical view of social responsibility is the ________________

a. stakeholder theory.
b. government subsidy theory.
c. taxing maxim.
d. "share your gain" maxim.
e. stockholder greed premise.

Answer: (a) Difficulty: (2)

45. An opposite view of the classical view of social responsibility is that business has a social responsibility to serve all ____________________ affected by its business decisions.

a. government programs
b. children
c. corporate shareholders
d. foreign aid programs
e. corporate stakeholders

Answer: (e) Difficulty: (1)

46. A ____________________ is any group which is vital to the survival and success of the corporation.

a. corporate stockholder
b. corporate advocate group
c. corporate board of directors
d. corporate stakeholder
e. corporate advisor

Answer: (d) Difficulty: (2)

47. All of the following have been traditionally identified as being stakeholders in a modern corporation EXCEPT ____________________

a. employees.
b. the community.
c. the government.
d. suppliers.
e. managers.

Answer: (c) Difficulty: (1)

48. Whereas the classical view of the corporation (with respect to social responsibility) focuses on maximizing its profits, stakeholder theory holds that the corporation should be managed __________

a. for the benefit of all its stakeholders.
b. for the benefit of the government and its needs.
c. for the benefit of international competition.
d. for the benefit of the underprivileged.
e. for the benefit of the few.

Answer: (a) Difficulty: (1)

49. Between the extremes of the classical theory and the stakeholder theory of social responsibility and corporate governance lies an intermediate position known as the________________

a. profit maxim.
b. moral minimum.
c. philosophy triangle.
d. Delphi approach.
e. greed-gain approach.

Answer: (b) Difficulty: (3)

50. Advocates of a middle ground between the classical theory and the stakeholder theory of social responsibility would say that the purpose of the corporation is to ________________

a. maximize profits within the law.
b. maximize profits, but share gain with the non-profit sector.
c. maximize profits, but subject to the requirement that it must do so in conformity with the moral minimum.
d. maximize profits so long as it is good for the societal tax base.
e. minimize expansion into questionable moral areas.

Answer: (c) Difficulty: (2)

51. The ________________ requirement advocated by those that would like to have a middle ground between classical theory and the stakeholder theory of social responsibility would say a firm should be free to strive for profits so long as it commits no harm.
a. profit maxim
b. moral minimum
c. philosophy triangle
d. Delphi approach
e. greed-gain approach

Answer: (b) Difficulty: (2)

52. Using the ________________ view toward corporate social responsibility, a business would certainly have a social responsibility to not produce an exploding cigarette lighter or poisonous chemical plants. However, it is unlikely that the social responsibilities of the business would extend beyond donating some of profits to charity or educating the poor (for example).

a. profit maxim
b. moral minimum
c. philosophy triangle
d. Delphi approach
e. greed-gain approach

Answer: (b) Difficulty: (2)

53. Given a commitment to being socially responsible, one way that a firm can measure their progress toward and commitment to social responsibility is to conduct a(n) ________________

a. social probe.
b. social responsibility audit.
c. ethical probe.
d. communications audit.
e. diversity sensitivity scenario.

Answer: (b) Difficulty: (1)

54. When employees try to report organizational wrongdoing, they are called ______________

a. traitors.
b. spies.
c. scabs.
d. 5th columnists.
e. whistle-blowers.

Answer: (e) Difficulty: (1)

55. ________________ means "planning and implementing organizational systems and practices to manage people so that the potential advantages of diversity are maximized while its potential disadvantages are minimized."

a. Marketing social responsibility
b. Preaching social responsibility
c. Managing diversity
d. Pursuing ethical conduct
e. Encouraging social responsibility

Answer: (c) Difficulty: (2)

56. A workforce is ______________ when it is comprised of two or more groups, each of whose members are identifiable and distinguishable based on demographic or other characteristics.

a. biased
b. ethnically oriented
c. stereotyped
d. racially sensitive
e. diverse

Answer: (e) Difficulty: (1)

57. When managers talk about "diversity," they usually are talking about one or all the groups listed below EXCEPT ____________________

a. racial and ethnic groups.
b. political groups or parties.
c. women.
d. older workers.
e. sexual/affectional orientation.

Answer: (b) Difficulty: (2)

58. The primary barriers in dealing with diversity are characterized as being all of the following EXCEPT ____________________

a. place of employment.
b. stereotyping and prejudices.
c. ethnocentrism.
d. discrimination.
e. tokenism.

Answer: (a) Difficulty: (1)

59. ____________________ is the process in which specific behavioral traits are ascribed to individuals on the basis of their apparent membership in a group.

a. Gender-role
b. Stereotyping
c. Ethnocentrism
d. Discrimination
e. Tokenism

Answer: (b) Difficulty: (2)

60. ____________________ is the tendency for viewing members of one's own group as the center of the universe and for viewing other social groups less favorably than one's own.

a. Gender-role
b. Stereotyping
c. Ethnocentrism
d. Discrimination
e. Tokenism

Answer: (c) Difficulty: (2)

61. ____________ refers to a behavioral bias toward or against the person based on the person's group.

a. Gender-role
b. Stereotyping
c. Ethnocentrism
d. Discrimination
e. Tokenism

Answer: (d) Difficulty: (2)

62. The "glass ceiling" enforced by the "old boys' network" would be an example of ____________

a. prejudices
b. stereotyping
c. ethnocentrism
d. discrimination
e. tokenism

Answer: (d) Difficulty: (3)

63. All of the following organizational activities are at the heart of managing any diversity program EXCEPT ____________

a. assessing the desire for advancement and increased pay for performance.
b. providing strong leadership.
c. researching: assessing the situation.
d. providing diversity training and education.
e. changing culture and management systems.

Answer: (a) Difficulty: (3)

64. For almost all companies that have made great strides in managing diversity, which of the following steps was not only crucial but had to be there for success to be achieved?

a. Assess the desire for advancement and increased pay for performance.
b. Provide strong leadership.
c. Research: assess the situation.
d. Provide diversity training and education.
e. Change culture and management systems.

Answer: (b) Difficulty: (3)

65. One of the most successful techniques for educating and training the workforce in diversity issues has been to adopt a policy of ________________. This is defined as being a relationship between a younger adult and an older, more experienced adult in which the older adult provides support, guidance, and counseling to enhance the younger adult's success at work and in other arenas of life.

a. teacher-student
b. boss-subordinate
c. counselor
d. mentoring
e. monitoring

Answer: (d) Difficulty: (2)

True/False and Short Answer

66. A normative judgment refers to the principles of conduct governing an individual or a group.

Answer: (False) Difficulty: (2)

67. ______________ refers to the principles of conduct governing an individual or a group.

Answer: (Ethics) Difficulty: (2)

68. A normative judgment implies that something is good or bad, right or wrong, better or worse.

Answer: (True) Difficulty: (1)

69. "That's a great looking car" is an example of a normative judgment.

Answer: (True) Difficulty: (1)

70. Ethical decisions always involve morality.

Answer: (True) Difficulty: (2)

71. The law itself is not an adequate guide to whether or not a decision is ethical.

Answer: (True) Difficulty: (2)

72. The manager shoulder, most of the credit (or blame) for any ethical decisions made by individuals on the job since the organization develops and shows what is ethical behavior.

Answer: (False) Difficulty: (2)

73. With the exception of years of business experience, studies have shown that characteristics (such as gender, age, marital status, education, dependent children, region of the country, and years in business) are not very good predictors of ethical behavior.

Answer: (False) Difficulty: (3)

74. It is hard to generalize about the characteristics of ethical or unethical people.

Answer: (True) Difficulty: (2)

75. Most people view themselves as being more ethical than others and, therefore, tend to have a distorted view of just how ethical they really are.

Answer: (True) Difficulty: (2)

76. Studies have shown that with respect to unethical behavior in organizations, ethical crises are most often caused by one unscrupulous employee.

Answer: (False) Difficulty: (3)

77. The Sears case presented in the text was an example of how unethical business practice involves tacit, if not explicit, cooperation of others and demonstrates the organizational culture.

Answer: (True) Difficulty: (2)

78. A ________________ is a basic belief about what you should and should not do, and what is and is not important.

Answer: (value) Difficulty: (3)

79. The Personnel Department's actions may be "the single most important factor in fostering corporate behavior of a high ethical standard."

Answer: (False) Difficulty: (2)

80. In the studies indicated in the text with regard to corporate ethics, no point emerges more clearly than the crucial role of top management.

Answer: (True) Difficulty: (1)

81. Corporate ________________ refers to the extent to which companies should and do channel resources toward improving one or more segments of society other than the firm's own stockholders.

Answer: (social responsibility) Difficulty: (2)

82. A ________________ is any group which is vital to the survival and success of the corporation.

Answer: (corporate stakeholder) Difficulty: (3)

83. Between the classical position and the stakeholder theory of social responsibility is an intermediate position known as the ____________________.

Answer: (moral minimum) Difficulty: (3)

84. Whistle-blowing describes the activities of employees who try to report organizational wrongdoing.

Answer: (True) Difficulty: (1)

85. ________________ refers to the tendency of some companies to appoint a relatively small group of women or minority-group members to high-profile positions rather than more aggressively achieving full work-group representation for that group.

Answer: (Tokenism) Difficulty: (3)

Essay

86. Ethical decisions always involve morality, in other words, society's accepted norms of behavior. It has been found that moral standards differ from non-moral ones in five main ways. Briefly, characterize each of those ways and give a brief example of each.

Answer:

a. Moral standards address matters of serious consequence to human well-being. Moral standards against murder, lying, slander, or price-fixing exist because there is a consensus that such actions pose a serious threat to individuals' or society's well-being.

b. Moral standards cannot be established or changed by decisions of authoritative bodies like legislatures. Instead, the question of what is morally right or wrong transcends the actions of authoritative bodies.

c. Moral standards should override self-interest. Thus, if you have a moral obligation to do something, you are expected to do it even if doing so might (for instance) impede your career or cost you money.

d. Moral judgments are never situational. Something that is morally right (or wrong) in one situation is right (or wrong) in another. For example, the fact that your company will benefit if you pay a bribe does not make the bribe right.

e. Moral judgments tend to trigger strong emotions. For example, violating moral standards may make you feel ashamed or remorseful. If you see someone else acting immorally, you may feel indignant or resentful.

Difficulty: (2)

87. In a study cited in your text, it was suggested that the nature of the moral issue (and its intensity) can influence the ethical decision the manager makes. In the study it was found that three rationales were used to justify the ethical judgments made. They were (1) the decision was made to avoid punishment; (2) the decision was made to uphold others' expectations; and (3) the decision was made to comply with moral conscience. Read the scenario described below, analyze it, make what you perceive to be an ethical decision, and justify your decision based on one of the three rationales cited above. Be sure to support your answer and conclusions.

Ethical scenario:

You are a student that is part of group of 20 students taking a test in Principles of Management. The test is being closely monitored by the instructor. A student worker comes to the classroom door and asks the instructor to step outside to receive a message. The instructor complies. When the instructor leaves, a student (whom you only casually know) takes out the exam answers from his notebook and quickly copies the answers and returns them before the instructor returns to class a few moments later. You are the only one to see the infraction, the exam will be graded on a curve, and you now have a dilemma. What do you do and how do you justify your action?

Answer:

Before assigning this exam question, it is recommended that the instructor review the material in the text and make a personal decision as to what answers might be acceptable to this question. According to material in the text, the answer could be justified in any of three areas. However, choice #3 is the strongest. Be sure to examine the justification for the answer. It might also be a good idea to use this material as a follow-up question (for discussion) purposes after the test to discuss various answers that were received. The instructor will also notice that a difference may be present between what one should do and what one would do.

Difficulty: (3)

88. With respect to studies cited in the text, summarize how to foster ethics at work. You may list and discuss. Be sure to cite what you perceive to be the most important factors or steps based on research (not just personal opinion).

Answer:

a. Emphasize top management's commitment. In the experience of the companies that were studied with regard to corporate ethics, no point emerges more clearly that the crucial role of top management in setting standards and enforcing policy.

b. Publish a "code." Firms with effective ethics programs set forth principles of conduct for the whole organization in the form of written documents.

c. Establish compliance mechanisms. Establish mechanisms to ensure compliance with the firm's ethical standards.

d. Involve personnel at all levels. Developing effective corporate ethics programs requires involving those whose behavior the ethics policies are to influence.

e. Measure results. Although the end results cannot all be precisely measured, all excellent firms monitor compliance with ethical standards.

Difficulty: (2)

89. Compare and contrast managerial capitalism, stakeholder theory, and the moral minimum. Be sure to explain each concept before embarking on your discussion.

Answer:

a. Managerial capitalism. The classical view is that a corporation's primary (and perhaps sole) purpose is to maximize profits for its stockholders. This view is presented by Milton Friedman. Friedman believes that "few trends could so thoroughly undermine the very foundation of our free society as the acceptance by corporate officials of a social responsibility other than to make as much money for their stockholders as possible.

b. Stakeholder theory. This is often presented as the opposite of the classical view. This view says that a business has the social responsibility to serve all "corporate stakeholders (any group which is vital to the survival and success of the corporation)." The benefits of wealth in business should benefit all.

c. The moral minimum. Between the extremes presented above, is the intermediate position of the moral minimum. Its advocates would agree that the purpose of the corporation is to maximize profits, but subject to the requirement that it must do so in conformity with the moral minimum (meaning that the firm should be free to strive for profits so long as it commits no harm).

Difficulty: (3)

90. Any attempt at managing diversity has to begin with an understanding of the barriers that may prevent a company from taking full advantage of the potential in its diverse workforce. List and briefly describe the more important of these barriers.

Answer:

a. Stereotyping and prejudices. Stereotyping is a process in which specific behavioral traits are ascribed to individuals on the basis of their apparent membership in a group. Prejudice is a bias that results from prejudging someone on the basis of some trait.

b. Ethnocentrism. This is a prejudice on a grand scale. It is the tendency for viewing members of one's own group as the center of the universe and for viewing other social groups (out-groups) less favorably than one's own.

c. Discrimination. Whereas prejudice means prejudging someone based on that person's traits, discrimination refers to a behavioral bias toward or against the person based on the person's group (such as age or gender).

d. Tokenism. Tokenism refers to the tendency of some companies to appoint a relatively small group of women or minority-group members to high-profile positions rather than more aggressively achieving full work-group representation for that group.

e. Gender roles. This is the tendency to associate women with less prestigious and non-managerial jobs.

Difficulty: (2)

Chapter 4
MAKING DECISIONS

INTRODUCTION

This chapter describes how managers make decisions that contribute to the achievement of organizational effectiveness. The decision-making process is analyzed, and different types of decisions are examined. Barriers to effective decision-making are identified, and the use of groups in the decision making process is analyzed.

LECTURE OUTLINE

What's Ahead?

Understanding Decision Making
- Decisions and the Management Process
- Programmed and Nonprogrammed Decisions
- The "Rational" Decision-Making Process

How Managers Make Decisions: The Limits to Rationality
- Individual Differences
- Decision-Making Shortcuts: Heuristics
- Decision Biases
- The Framing of the Problem
- Escalation of Commitment
- Functional Fixedness
- Psychological Set
- Organizational Decision-Making Barriers

Decision Making in Practice: Bounded Rationality
- Decision Making: How It Should Be Versus How It Actually Is
- Satisficing Versus Optimizing

How to Make Better Decisions
- Increase Your Knowledge
- Debias Your Judgment
- Be Creative
- Use Your Intuition
- Don't Overstress the Finality of Your Decision
- Be Sure the Timing Is Right

Using Groups to Make Better Decisions
- Groups at Work
- Pros and Cons of Group Decision Making
- Tools for Improving Group Decision Making

Summary

Case: Belly-Up at Palm Island Bank ?
You Be the Consultant: Identifying Top Talent

ANNOTATED OUTLINE

What's Ahead

Micro Vision has some important decisions to make regarding the marketing of its new way to display computer information. Decision making is the process of developing and analyzing alternatives and making a choice. Decision making is often the consequence of needing to solve a problem. And, decision making is often subject to personal distortions and biases, making it a judgmental process.

Notes:

Understanding Decision Making

Decisions and the Management Process. Decision making is a key to effective management. Each of the four primary management functions requires effective decision making. Nearly everything a manager does requires decision making.

Notes:

Programmed and Nonprogrammed Decisions. Any decision made by a manager can be classified as either programmed or nonprogrammed. Programmed decisions are repetitive, routine, and can be solved through mechanical procedures such as applying rules. It is estimated that up to 90 percent of decisions are programmed. Nonprogrammed decisions are unique and novel, and mechanical procedures are not available. Crisis decisions are an example of nonprogrammed decisions.

Notes:

The "Rational" Decision-Making Process. The rational decision-making process includes defining the problem, identifying criteria for decision making, weighing the criteria by importance, developing alternatives, and analyzing alternatives. The decision maker then makes a decision, implements it, and evaluates it subsequently. In practice, decision making usually cannot be done as accurately or rationally as the decision maker would prefer. There are many factors that limit just how rational any decision maker can be.

Notes:

How Managers Make Decisions: The Limits to Rationality

Individual Differences. An individual's ability to absorb, analyze, and process information is limited. Most people reach information overload quickly and adjust in various ways. Decision making is further complicated by the fact that much information is lost when a decision goes through several levels of management.

Notes:

Decision-Making Shortcuts: Heuristics. People often take shortcuts, or heuristics, by applying rules of thumb to help them deal with the complexities of decision making. This undermines the ideal of the perfectly rational decision maker. Either the similarity heuristic or the availability heuristic may be applied by decision makers to simplify their efforts. The similarity heuristic is used when a manager tries to predict a person's performance based on that person's similarity to other individuals he has known in the past. The availability heuristic is applied when the decision maker uses more recent information that is more readily available in memory to make a decision.

Notes:

Decision Biases. Person-specific and cognitive biases can distort decision making. Person-specific biases are specific to an individual. Cognitive biases are defined as standard errors in reaching a decision that may occur, for example, when a decision maker is overconfident and does not evaluate alternatives effectively. Another problem is the cognitive trap where decision makers tend to seek only information that confirms what they believe is true and neglect to search for disconfirming information.

Notes:

The Framing of the Problem. The way the problem is presented can have a significant influence on the rationality of the decision. If the problem is framed incorrectly, a decision maker could make a significant error.

Notes:

Escalation of Commitment. Escalation of commitment occurs when each side becomes more committed to its position when decisions have to be made step by step.

Notes:

Functional Fixedness. The inability to see that an object can have a function other than its stated or usual one is functional fixedness. This approach can limit a manager's ability to discover alternative solutions to problems.

Notes:

Psychological Set. The tendency to focus on a rigid strategy or approach to certain types of problems is called psychological set. This approach also limits a manager's ability to find alternative solutions.

Notes:

Organizational Decision-Making Barriers. The organization, itself, can sometimes impede successful decision making. For example, subordinates are often reluctant to bring bad news to the attention of their superiors. This lack of communication can result in bad decisions for the organization.

Notes:

Decision Making in Practice: Bounded Rationality

Decision Making: How It Should Be Versus How It Actually Is. The decision-making barriers identified above illustrate that decision making is rarely performed the way it "should be" and is more often modified to account for how managers actually make decisions. A manager's decision making is only as rational as his or her unique values, capabilities, and limited capacity for processing information permit him or her to be.

Notes:

Satisficing Versus Optimizing. While the rational and the bounded rational decision maker continue to review solutions until they find the optimal choice, in practice, managers often have to satisfice by selecting satisfactory alternatives, not necessarily optimal alternatives. For example, the costs of seeking optimal solutions may be excessive.

Notes:

How to Make Better Decisions

Increase Your Knowledge. As decision makers gain experience, they tend to be more effective in making decisions. Additionally, the more experience the decision maker has, the more confident he or she will be in making an informed choice.

Notes:

Debias Your Judgment. If decision makers reduce or eliminate biases from their judgmental processes, they can be more effective in reaching good decisions. Decision makers need to understand that the possibility of bias exists, understand how the bias can affect their judgment, analyze previous decisions to provide feedback on where bias may have influenced judgment, and accept the fact that biases exist but can be reduced or eliminated through diligent efforts.

Notes:

Be Creative. An essential aspect of successful decision making is creativity--the process of developing original, novel responses to a problem. Creativity is needed to develop new alternatives and correctly define the problem. Decision makers should check their assumptions and think through the whole decision-making process to improve creativity.

Notes:

Use Your Intuition. Many behavioral scientists have argued that overemphasizing rationality and logic can actually backfire by blocking the decision maker from using intuition. Intuition is a cognitive process whereby a person unconsciously makes a decision based on his or her accumulated knowledge and experience.

Notes:

Don't Overstress the Finality of Your Decision. Decision makers need to recognize that very few decisions are forever--it is not necessary to become frozen in the finality of a decision. It is not only possible, it is smart to correct a poor decision.

Notes:

Make Sure the Timing Is Right. Decision makers need to recognize that their mood at the time could have a significant impact on the quality of the decision. Decision makers should not be swayed by passing moods. One research study found that decision makers tend to be aggressive and destructive when making decisions when they feel "down," and tolerant and balanced in decision making when they feel good.

Notes:

Using Groups to Make Better Decisions

Groups at Work. A group is two or more persons who interact together for some purpose and in such a manner that each person influences and is influenced by each other person. The norms and cohesiveness of the group can have a significant impact upon individual behavior and group performance.

Notes:

Pros and Cons of Group Decision Making. Groups can make higher-quality decisions because of the wider range of experience and knowledge employed in reaching the decision. Also, if the group formulates its own decision, it is likely to be more committed to successful implementation of the decision. However, problems can occur. People may not question an issue to be considered for fear of not being considered a good group member. Also, a dominant individual may emerge and control the participation of certain group members. Group decision making can be time consuming and more costly. Group members may disagree on issues and spend too much time trying to satisfy their egos by winning an argument rather than trying to solve a problem. Groupthink may occur when a group becomes too cohesive and people do not challenge a decision even though they believe it is wrong.

Notes:

Tools for Improving Group Decision Making. There are several effective tools available for improving group decision making. Brainstorming is a way to use the creative energies of the group members by introducing all possible solutions before evaluating any of them. Participants are expected to avoid criticizing others' ideas, share even wild suggestions, offer as many suggestions and supportive comments as possible, and build on others' suggestions to create their own. The devil's-advocate approach can improve decision making by requiring people to examine the alternatives more thoroughly. The Delphi technique obtains the opinions of experts who work independently with the experts' opinions from one stage providing the basis for the experts' analyses of each succeeding stage. Several rounds are conducted until a consensus is reached. The nominal group technique attempts to minimize the probability of problems like groupthink. Members write down their ideas, present them orally, discussion is held about the ideas presented by participants, secret votes are held on proposals, and those with the most votes are selected. The stepladder technique includes adding group members at each stage in the decision-making process so each participant's input is untainted by input of his or her predecessors.

Persons leading group decision making must allow all members to participate, distinguish between idea getting and idea evaluation, not attempt to dominate the discussion, and ensure that effort is directed toward overcoming obstacles.

Notes:

Summary

Notes:

Case: Belly-Up at Palm Island Bank ?

1. The president should be very concerned about escalation of commitment. Should she spend more to bail out an already sinking ship?

2. The president needs to take a bounded rational approach. If a purely rational approach is taken, the president may decide to spend another $250,000 rather, absorb a virtually certain original loss of $250,000--escalation of commitment may not be recognized.

3. Perhaps the bank could tie a new $250,000 loan to the accounts receivable of the family owned business and the jeans suppliers to ensure recovery of the money. Also, perhaps a deal could be negotiated with the suppliers of the family owned business to let them pay when they are paid by the jeans suppliers.

4. The bank should not loan more money unless there is sufficient, additional collateral to guarantee repayment or the suppliers of the family owned business agree to extend payment deadlines to coincide with receipt of revenue from the jeans suppliers.

You Be the Consultant: Identifying Top Talent

1. The firm could use brainstorming, the Delphi technique, or the nominal group technique for creativity-stimulating techniques. They should also check their assumptions and think thoroughly through the process. Intuition should be part of the process and they need to recognize that decisions are based on bounded rationality.

2. KnitMedia should take advantage of its multicultural workforce and use group decision making and creativity. The managers come from several different countries, and each of them should have interesting approaches for attracting talent in their home areas. They should be able to come up with some excellent, specific ideas for proceeding in each talent market where their managers have first hand knowledge about the country.

End of Chapter Questions for Chapter 4

Multiple Choice

1. A(n) ______________ is a choice from among available alternatives.

a. heuristic
b. code diagram
c. decision
d. rational process diagram
e. intuition

Answer: (c) Difficulty: (1)

2. A decision is ________________

a. the process of developing and analyzing alternatives and making a choice.
b. a discrepancy between a desirable and an actual situation.
c. a reference to the cognitive or "thinking" aspects of the decision-making process.
d. a choice from among the available alternatives.
e. repetitive, routine, and usually solved through mechanical procedures such as applying rules.

Answer: (d) Difficulty: (3)

3. ______________ is the process of developing and analyzing alternatives and making a choice.

a. Heuristic modeling
b. Code diagraming
c. Decision making
d. Rational process diagraming
e. Intuitive thinking

Answer: (c) Difficulty: (2)

4. Decision making is ______________________

a. the process of developing and analyzing alternatives and making a choice.
b. a discrepancy between a desirable and an actual situation.
c. a reference to the cognitive or "thinking" aspects of the decision-making process.
d. a choice from among the available alternatives.
e. repetitive, routine, and usually solved through mechanical procedures such as applying rules.

Answer: (a) Difficulty: (3)

5. Most decisions are prompted by problems. A problem is ___________________

a. the process of developing and analyzing alternatives and making a choice.
b. a discrepancy between a desirable and an actual situation.
c. a reference to the cognitive or "thinking" aspects of the decision-making process.
d. a choice from among the available alternatives.
e. repetitive and routine and usually solved through mechanical procedures such as applying rules.

Answer: (b) Difficulty: (2)

6. A ________________ refers to the cognitive or "thinking" aspects of the decision-making process.

a. problem
b. decision
c. heuristic
d. nominal technique
e. judgment

Answer: (e) Difficulty: (2)

7. Which of the following statements about the relationship between a "problem" and a "decision" is most <u>TRUE</u> from the choices available?

a. A decision always involves a problem or a problem situation.
b. A decision doesn't necessarily involve a problem.
c. A decision is the same thing as a judgment.
d. Decision making is usually free of distortions and biases.
e. Decision making and problem solving are often not the same thing since they usually involve different cognitive skills.

Answer: (b) Difficulty: (3)

8. Consider the following typical decisions that managers face: (a) What are the organization's long-term objectives? (b) What strategies will best achieve these objectives? (c) What should the organization's short-term objectives be?

   The typical managerial decisions listed above would most likely be part of which of the following management function categories?

a. planning
b. organizing
c. leading
d. controlling
e. networking

Answer: (a) Difficulty: (2)

9. Consider the following typical decisions that managers face: (a) How do I handle employees who appear to be low in motivation? (b) How will a specific change affect worker productivity? (c) When is the right time to stimulate conflict?

   The typical managerial decisions listed above would most likely be part of which of the following management function categories?

a. planning
b. organizing
c. leading
d. controlling
e. networking

Answer: (c) Difficulty: (2)

10. ______________ decisions are decisions that are repetitive and routine and that can be solved through mechanical procedures such as by applying decision rules.

a. Choice
b. Nonprogrammed
c. Programmed
d. Intuitive
e. Rational

Answer: (c) Difficulty: (2)

11. "If a customer returns a jacket, you may give that person a refund if the tag is not removed, if the jacket is not damaged, and if the purchase was made within the past two weeks," is an example of which of the following types of decisions?

a. Choice
b. Nonprogrammed
c. Programmed
d. Intuitive
e. Rational

Answer: (c) Difficulty: (2)

12. Some writers estimate that as much as 90 percent of all management decisions are of the ______________ variety.

a. Delphi
b. nonprogrammed
c. intuitive
d. rational
e. programmed

Answer: (e) Difficulty: (3)

13. Which of the following decision-making forms are considered to be novel, unique, and do not usually have mechanical procedures for making them?

a. Delphi
b. nonprogrammed
c. programmed
d. rational
e. intuitive

Answer: (b) Difficulty: (1)

14. Crisis decisions (like managing the rescue work for a plane crash) are examples of ____________

a. programmed decision making.
b. employee-based rational decision making.
c. nonprogrammed decision making.
d. irrational decision making.
e. rational decision making.

Answer: (c) Difficulty: (2)

15. Generally speaking, ____________________ decisions rely heavily on judgment and focus on the firm's long-term strategic development and survival.

a. programmed
b. irrational
c. nonprogrammed
d. cognitive
e. rational

Answer: (c) Difficulty: (3)

16. If a decision is seen to be procedural, predictable, and has well-defined information and decision criteria, it is probably a(n) ________________________ decision.

a. programmed
b. irrational
c. nonprogrammed
d. cognitive
e. rational

Answer: (a) Difficulty: (3)

17. Top managers tend to confront more unstructured, ambiguous situations (such as "What will our competitors do?). These situations require more ____________ decisions to be made.

a. programmed
b. energy-charged
c. nonprogrammed
d. cognitive
e. rational

Answer: (c) Difficulty: (2)

18. "How many items should we produce today?" would normally be considered to be which of the following decision forms?

a. programmed
b. employee biased
c. nonprogrammed
d. cognitive
e. rational

Answer: (a) Difficulty: (2)

19. All of the following are assumptions about a "rational manager" EXCEPT ______________

a. the rational manager had complete or "perfect" information about the situation including the full range of goods and services available on the market and the price of each good or service.
b. the rational manager could define the problem and not get confused by symptoms or other obstacles.
c. the rational manager would always choose the path that led to maximum profit.
d. the rational manager could identify all criteria and accurately weight all the criteria according to their preferences.
e. the rational manager knew all the possible alternatives and could accurately assess each against each criterion.

Answer: (c) Difficulty: (3)

20. The "rational" manager's approach to making a decision would include all of the following steps EXCEPT ______________

a. defining the problem.
b. identifying the crisis environment.
c. identifying the criteria.
d. analyzing the alternatives.
e. selecting an alternative.

Answer: (b) Difficulty: (1)

21. Your firm wants to buy a computer to assist employees with their work load. Through focus-group analysis, employees have revealed that price, the manufacturer and model of the computer, warranties, service support, reliability, and repair record of the computer would all be areas that were important to the workers and managers that might use the computers. Now that these criteria have been revealed, the decision makers are ready to proceed to which of the following steps in the decision-making process?

a. Identify the criteria.
b. Develop alternatives.
c. Analyze alternatives.
d. Weight the criteria.
e. Choose an alternative.

Answer: (d) Difficulty: (1)

22. With respect to the "rational" manager's approach to making a decision, an expert has characterized the ________________ step as being "the most difficult part of the decision-making process, because this is the stage that typically requires forecasting future events."

a. "identify the criteria"
b. "develop alternatives"
c. "analyze alternatives"
d. "weight the criteria"
e. "choose an alternative"

Answer: (c) Difficulty: (3)

23. In practice, many factors limit just how rational any decision maker can be. All of the following are significant barriers or limitations to making rational decisions <u>EXCEPT</u> (pick the best response) ________________

a. individual differences.
b. profit maximization.
c. decision-making shortcuts (heuristics).
d. functional fixedness.
e. psychological set.

Answer: (b) Difficulty: (2)

24. In one series of lab studies, subjects were required to make decisions based on the amount of information transmitted on a screen. Most people quickly reached a point at which "information overload" occurred, and they began adjusting in several ways (such as omitting or ignoring information transmitted on the screen). These errors were the result of a barrier to rational decision making. Which of the responses below MOST accurately describes the type of barrier that the text identifies as being the cause of the failure to achieve rational decisions in the above illustration?

a. individual differences
b. decision biases
c. decision-making shortcuts (heuristics)
d. functional fixedness
e. psychological set

Answer: (a) Difficulty: (3)

25. ________________ is the selection and interpretation of information we receive through our senses and the meaning we give to the information.

a. Judgment
b. Sensitivity
c. Perception
d. Choice determination
e. Modeling

Answer: (c) Difficulty: (1)

26. In organizations, a manager's prior experiences and position in the company can have a big effect on how the manager sees a problem and how he or she reacts to it. Which of the following terms <u>BEST</u> describes the area that impacts the above statement?

a. Judgment
b. Sensitivity
c. Perception
d. Choice determination
e. Modeling

Answer: (c) Difficulty: (3)

27. Apart from (or perhaps because of) differences in perception and ability, individuals differ in their approaches to decision making. People who are ______________ decision makers tend to take a more logical, structured, step-by-step approach to solving a problem.

a. systematic
b. intuitive
c. biased
d. ethnocentric
e. dynamic

Answer: (a) Difficulty: (1)

28. ________________ (decision-making shortcuts or rules of thumb) undermine the economist's ideal of a perfectly rational, optimizing decision maker.

a. Judgments
b. Irrationality
c. Intuition
d. Framing
e. Heuristics

Answer: (e) Difficulty: (2)

29. If a banker states that "people should only spend 30 percent of their disposable income for a mortgage (plus interest expenses)," the banker is using a(n) __________ to set policy. This could easily undermine the rationality of the banker's decisions.

a. judgment
b. irrational thought
c. intuition
d. heuristic
e. bias

Answer: (d) Difficulty: (2)

30. A common problem in appraising a person's performance is to evaluate the person based mostly on recent performance (the last few weeks rather than the whole year) because those experiences are recalled from memory. This statement illustrates the use of which of the following decision-making shortcuts?

a. a similarity heuristic
b. a choice heuristic
c. an availability heuristic
d. a bias heuristic
e. a transference

Answer: (c) Difficulty: (3)

31. Several types of bias can distort the decision a person makes. The ____________ bias is a bias specific to particular people or groups.

a. gender
b. foreign
c. person-specific
d. xenophobic
e. color

Answer: (c) Difficulty: (2)

32. Someone biased in favor of a particular ethnic group might consciously or unconsciously decide to hire more people from that group. This form of bias is called

a. gender bias.
b. foreign bias.
c. xenophobic bias.
d. person-specific bias.
e. color bias.

Answer: (d) Difficulty: (2)

33. Most people are influenced by what psychologists call ____________ biases, which are defined as standard errors in reaching a decision.

a. gender
b. cognitive
c. recognition
d. processing
e. person-specific

Answer: (b) Difficulty: (3)

34. People tend to be overconfident and to overestimate their ability in answering questions with which they are quite familiar. The problem is they do not correspondingly reduce their confidence level when solving relatively difficult problems that are outside their areas of expertise. The above statements are examples of what psychologists call ________________ biases.

a. gender
b. cognitive
c. recognition
d. processing
e. person-specific

Answer: (b) Difficulty: (3)

35. Consider the following situation: Suppose you have tentatively decided to buy a new computer. If you are like most people, your research now will tend to focus more on information that supports your initial decision than on information that does not support it. Like most people, you are the unwitting victim of a cognitive decision bias form called __________________

a. the Delphi illusion.
b. the first-in parallel.
c. the "want-it" syndrome.
d. the "quick decision" failure.
e. the "confirmation" trap.

Answer: (e) Difficulty: (3)

36. A company president is told that the following choice exists: Settle a lawsuit out of court for $10 million and thus accept a sure loss of $10 million, or go to court and have a 50 percent chance of a $20 million loss. When presented with a problem that is described in this manner, most people would choose to go to court. However, presented differently, the option could be stated as settle out of court for $10 million and thus save $10 million, or go to court and have a 50 percent chance of losing $20 million. This dilemma illustrates that ___________ can significantly influence the rationality of the decision.

a. the amount of money involved
b. threat involved in a problem
c. the framing of the problem
d. the perception of the significance
e. the futility of the problem

Answer: (c) Difficulty: (3)

37. Have you ever had to negotiate with someone only to find that both of your positions became more and more entrenched the more you negotiated? This problem is an illustration of a difficulty in rational decision making called ______________________
a. functional fixedness.
b. psychological set.
c. decision biases.
d. framing the problem.
e. escalation of commitment.

Answer: (e) Difficulty: (2)

38. ___________ "is the inability to see that an object can have a function other than its stated or usual one." This decision-making barrier can limit a manager's ability to discover alternative solutions.

a. Functional fixedness
b. A psychological set
c. A decision bias
d. Framing the problem
e. Escalation of commitment

Answer: (a) Difficulty: (1)

39. A mental trait that can severely limit a manager's ability to find alternative solutions is called ______________. This is the tendency to focus on a rigid strategy or approach to certain types of problems.

a. functional fixedness
b. psychological set
c. decision bias
d. framing the problem
e. escalation of commitment

Answer: (b) Difficulty: (2)

40. Consider the following problem: In one company, lower-level managers knew a product was in serious trouble. However, they were reluctant to tell their superiors because they knew their superiors had done the marketing and production studies that led to the decision to produce the product. They knew their bosses would resist their information and be critical of the person(s) bearing the bad news. This is an example a barrier to rational decision making generally known as ______________

a. the Delphi illusion.
b. organizational decision-making barriers.
c. the "want-it" syndrome.
d. the "quick decision" failure.
e. the "confirmation" trap.

Answer: (b) Difficulty: (2)

41. ______________ means that a manager's decision making is only as rational as his or her unique values, capabilities, and limited capacity for processing information permit him or her to be.

a. The Delphi illusion
b. Organizational decision-making barriers
c. Rationality
d. Bounded rationality
e. The "confirmation" trap

Answer: (d) Difficulty: (2)

42. Consider the following statements: (a) The decision maker lacks information to completely define the problem; (b) The decision maker may not be able to knowledgeably weigh all criteria; and (c) The decision maker hasn't the time or money to assess all the alternatives completely. All of the above illustrate the concept of ______________.

a. the Delphi illusion.
b. organizational decision-making barriers.
c. rationality.
d. bounded rationality.
e. the "confirmation" trap.

Answer: (d) Difficulty: (2)

43. Many managers are pressed by time and circumstances when attempting to make decisions. This can lead to problems in judgment. To ______________ means that managers in practice tend to be concerned with just discovering and selecting satisfactory alternatives, and only in exceptional cases with finding optimal alternatives.

a. rationalize
b. project
c. satisfice
d. sub-optimize
e. sublimate

Answer: (c) Difficulty: (1)

44. All of the following are ways to make better decisions EXCEPT to ______________

a. increase your knowledge.
b. tie decision making to money.
c. de-bias your judgment.
d. be creative.
e. use your intuition.

Answer: (b) Difficulty: (1)

45. All of the following are steps to de-bias your judgment EXCEPT to ______________

a. deal with the bias by incorporating it into decision making.
b. understand that the possibility of bias exists.
c. understand how bias can affect your judgment.
d. analyze previous decisions made to give yourself feedback on whether and how bias has influenced your judgment.
e. accept the fact that such biases exist but can be reduced or eliminated through your diligent efforts.

Answer: (a) Difficulty: (2)

46. One of the keys to being creative in decision making is to always (pick the best answer) __________

a. use imagination.
b. check your assumptions
c. change the difficulty level of the assignment.
d. involve "groupthink."
e. understand how to communicate creatively.

Answer: (b) Difficulty: (3)

47. One way to boost creativity is to think your way through the ______________ of decision making. Consider what is involved in going from step one to the finish.

a. function
b. joy
c. process
d. scheme
e. level

Answer: (c) Difficulty: (1)

48. ________________ can be defined as a cognitive process whereby we unconsciously make a decision based on our accumulated knowledge and experiences.

a. Creativity
b. Stimulation
c. Perception
d. Intuition
e. Invention

Answer: (d) Difficulty: (2)

49. A(n) ______________ is defined as two or more persons who interact together for some purpose and in such a manner that each person influences and is influenced by each other person.

a. cabal
b. assembly
c. leadership association
d. group
e. unit

Answer: (d) Difficulty: (1)

50. Group __________________ is (are) defined as being the attraction of the group for its members.

a. cohesiveness
b. norms
c. functions
d. parallels
e. systems

Answer: (a) Difficulty: (2)

51. The informal rules that groups adopt to regulate and regularize group members' behavior are called __________________

a. cohesiveness.
b. norms.
c. functions.
d. parallels.
e. systems.

Answer: (b) Difficulty: (2)

52. All of the following are advantages of using groups to make decisions EXCEPT __________

a. "two heads are better than one."
b. pooling experiences and points of view of several people can lead to more points of view regarding how to define the problem.
c. groups that analyze a problem and come up with their own decisions also tend to "buy" into those decisions.
d. group decision making is usually quicker than individual decision making because of consensus.
e. groups work harder to implement their decisions because they have accepted the decision.

Answer: (d) Difficulty: (3)

53. All of the following are disadvantages of using groups to make decisions EXCEPT ________

a. rather than "two heads are better than one," critics believe that using groups can short-circuit the decision process.
b. groups usually do not work as hard to implement their decisions as do individuals.
c. silent disagreement among members (leading to acceptance of the decision) can reduce creative decisions.
d. a dominant individual can cut off debate on issues.
e. groups take longer to arrive at decisions than do individuals.

Answer: (b) Difficulty: (3)

54. ________________ has been defined as "a mode of thinking that people engage in when they are deeply involved in a cohesive group, when the members' desire for unanimity overrides their personal motivation to realistically appraise alternative courses of action."

a. Group cohesion
b. Group stress
c. Groupthink
d. Group cancellation
e. Groupkill

Answer: (c) Difficulty: (2)

55. Which of the following illustrations is most often cited as the classic illustration of "groupthink" and the problems associated with it?

a. a faculty member being brought up on charges for preaching racial disharmony
b. the Kennedy Administration's decision to invade Cuba at the Bay of Pigs
c. the Federal Reserve's decision to raise interest rates during a time of prosperity
d. a professor's decision to throw out a curve on a test (even though one was promised) because of a student's complaint about fairness
e. a minister's decision to ask for a church member's resignation because of differences of opinion on church doctrine

Answer: (b) Difficulty: (3)

56. All of the following would be statements that would be associated with the negative aspects of "Groupthink" EXCEPT ________________

a. "Don't ask, don't question."
b. "You must conform."
c. "We're on a mission."
d. "Express your opinion and then let's vote."
e. "We are Masters of the world."

Answer: (d) Difficulty: (2)

57. A list of warning signs that "groupthink" is in progress would include all of the following EXCEPT ________________

a. the illusion of invulnerability.
b. collective rationalization.
c. unquestioned morality.
d. negative stereotyping.
e. resistance to unanimity.

Answer: (e) Difficulty: (2)

58. One of the common problems associated with "groupthink" is that group members cast outsiders in negative terms, making it easier to ignore them because their opposition is expected. This is commonly called ________________

a. the illusion of invulnerability.
b. collective rationalization.
c. unquestioned morality.
d. negative stereotyping.
e. resistance to unanimity.

Answer: (d) Difficulty: (2)

59. One of the common problems associated with "groupthink" is that group members ignore or discredit warning signals that are contrary to group thinking. This is commonly called

a. the illusion of invulnerability.
b. collective rationalization.
c. unquestioned morality.
d. negative stereotyping.
e. resistance to unanimity.

Answer: (b) Difficulty: (2)

60. ________________ is one way to amplify the creative energies of a group. It has been defined as a group problem-solving technique whereby group members introduce all possible solutions before evaluating any of them.

a. Group decision making
b. Groupthink
c. Brainstorming
d. Consensus building
e. Problem solving

Answer: (c) Difficulty: (1)

61. All of the following are rules of brainstorming EXCEPT ________________

a. stay on task and do not deviate from the assigned mission focus.
b. avoid criticizing other's ideas.
c. share even wild suggestions.
d. offer as many suggestions and supportive comments as possible.
e. build on others' suggestions to create your own.

Answer: (a) Difficulty: (2)

62. Consider the following situation: A company purchases a new software package (called groupware) that allows group members to interact via the groupware with their PCs instead of face-to-face. Doing so results in a relatively large increase in the number of high-quality ideas generated by the group. Which of the following BEST describes the process the company is using?

a. groupthink
b. the Delphi Technique
c. brainstorming
d. the Nominal Group Technique
e. the Stepladder Technique

Answer: (c) Difficulty: (3)

63. In ____________________, you obtain the opinions of experts who work independently, with the experts' written opinions from one stage providing the basis for the experts' analyses of each succeeding stage.

a. the Devil's-Advocate Approach
b. the Delphi Technique
c. brainstorming
d. the Nominal Group Technique
e. the Stepladder Technique

Answer: (b) Difficulty: (3)

64. A group decision-making process in which participants do not attempt to agree as a group on any solution, but rather meet face to face and vote on all the solutions proposed after privately ranking the proposals in order of their preference is called ____________________

a. the Devil's-Advocate Approach
b. the Delphi Technique
c. brainstorming
d. the Nominal Group Technique
e. the Stepladder Technique

Answer: (d) Difficulty: (3)

65. A group decision-making process where group members are added one by one at each stage of the decision-making process so that their input is untainted by the previous discussant's points of view is called ________________

a. the Devil's-Advocate Approach
b. the Delphi Technique
c. brainstorming
d. the Nominal Group Technique
e. the Stepladder Technique

Answer: (e) Difficulty: (2)

66. How to lead a group decision-making discussion is not an easy task. According to a classic series of group decision-making studies, an effective discussion leader has the responsibility to do all of the following EXCEPT ____________________

a. see that all members participate.
b. agitate the group so that a variety of opinions will come out.
c. distinguish between idea getting and idea evaluation.
d. do not respond to each participant or dominate the discussion.
e. see that effort is directed toward overcoming surmountable obstacles.

Answer: (b) Difficulty: (3)

Questions from the Appendix to Chapter 4

67. In financial analysis, the ______________ is that volume of sales at which revenues just equal expenses. Here you have neither a profit nor a loss.

a. gross profit projection
b. break-even point
c. demand curve apex
d. slope of the cost curve
e. high point of the sales curve

Answer: (b) Difficulty: (1)

68. ________________ is a financial analysis decision-making aid that enables a manager to determine whether a particular volume of sales will result in losses or profits.

a. Break-even analysis
b. Linear programming
c. The Delphi Technique
d. A Waiting Line/Queuing Technique
e. A Decision Tree

Answer: (a) Difficulty: (2)

69. Break-even analysis uses all of the following basic concepts EXCEPT ______________

a. fixed costs.
b. variable costs.
c. owner's equity.
d. revenues.
e. profits.

Answer: (c) Difficulty: (2)

70. The break-even point is the point where ____________________

a. profit is maximized.
b. total revenue crosses the total costs line.
c. expenses are maximized.
d. fixed costs exceed variable costs.
e. demand becomes less relevant.

Answer: (b) Difficulty: (3)

71. Using the break-even formula:

```
P(X)  = F + V(X)where
    F = fixed costs
    V = variable costs per unit
    X = volume of output (in units)
    P = price per unit
```

Solve the following (state the answer in units):
F = fixed costs of $2,000.00
V = variable costs per unit of $1.75
P = $2.00 per unit.

a. 1,000 units
b. 2,000 units
c. 4,000 units
d. 8,000 units
e. 12,000 units

Answer: (d) Difficulty: (2)

72. ________________ is a mathematical method used to solve resource allocation problems.

a. Break-even analysis
b. Linear programming
c. The Delphi Technique
d. A Waiting Line/Queuing Technique
e. A Decision Tree

Answer: (b) Difficulty: (2)

73. Bank managers need to know how many tellers they should have so they will be able to meet the demands of customers (but not have too many and waste money). One decision-making technique that can be used to determine the number needed would be to use ________________________.

a. break-even analysis
b. linear programming
c. the Delphi Technique
d. a Waiting Line/Queuing Technique
e. a Decision Tree

Answer: (d) Difficulty: (2)

74. ______________________ are (is) used to solve problems for which information is incomplete or uncertain.

a. Statistical decision theory techniques
b. Linear programming
c. The Delphi Technique
d. A Waiting Line/Queuing Technique
e. A Decision Tree

Answer: (a) Difficulty: (3)

75. Statistical decision theory is based on the idea that a manager may face three degrees of uncertainty in making a decision. The three degrees are described as being ________________

a. risk, risk avoidance, and parallelism.
b. certainty, uncertainty, and reduction strategies.
c. certainty, risk, and avoidance.
d. certainty, uncertainty, and risk.
e. certainty, uncertainty, and loss.

Answer: (d) Difficulty: (3)

True/False Short Answer

76. Most decisions are prompted by problems.

Answer: (True) Difficulty: (1)

77. Programming is the process of developing and analyzing alternatives and making a choice.

Answer: (False) Difficulty: (2)

78. A decision is a discrepancy between a desirable and an actual situation.

Answer: (False) Difficulty: (2)

79. ________________ refers to the cognitive or "thinking" aspects of the decision-making process.

Answer: (Judgment) Difficulty: (3)

80. "If a customer buys music on a CD, has not opened the plastic wrap or broken the security seal, and returns the CD within three days of the receipt of the disk, then the customer will receive a full refund or another CD of equal value." This statement would be an example of a non-programmed decision.

Answer: (False) Difficulty: (2)

81. The advent of computers has dramatically boosted the number of organizational decisions that can now be "programmed."

Answer: (True) Difficulty: (1)

82. Decisions that are routine, generic, and computational in nature are called ________________ decisions.

Answer: (programmed) Difficulty: (3)

83. If the nature of a decision is procedural, predictable, and has well-defined information and decision criteria, then a programmed decision is most likely.

Answer: (True) Difficulty: (2)

84. If the nature of a decision is novel, unstructured, has incomplete channels of information, and unknown criteria, this decision is usually a ______________ decision.

Answer: (non programmed) Difficulty: (3)

85. Top-level managers tend to be confronted mostly by programmed decisions.

Answer: (False) Difficulty: (2)

86. When a manager is confronted with a situation that demands that a decision be reached, the "rational" manager's approach to making a decision usually includes a six-step process. The first step is to ________________.

Answer: (define the problem) Difficulty: (2)

87. A consulting team has been retained by the owners of a large office building. The office workers are upset because they have to wait what they consider to be a long time for an elevator to pick them up. Many are threatening to move out of the building. The owners believe that the problem was that the elevators were running too slowly. Is this the real problem? Answer True if you think the owners have correctly identified the problem. Answer False if you believe they have identified a symptom.

Answer: (False) Difficulty: (2)

88. A person's ability to absorb, analyze, and generally "process" information, while varying from person to person, is quite limited.

Answer: (True) Difficulty: (2)

89. ______________ are decision-making shortcuts (also called rules of thumb).

Answer: (Heuristics) Difficulty: (3)

90. The similarity heuristic is where a manager might base a decision on the aspects of the situation that are most readily "available" in memory.

Answer: (False) Difficulty: (2)

91. The framing trap occurs when people tend to seek confirmatory information for what they think is true and neglect to search for discomfirmatory evidence.

Answer: (False) Difficulty: (3)

92. ______________ has been defined as a "mode of thinking that people engage in when they are deeply involved in a cohesive group, when the members' desire for unanimity overrides their personal motivation to realistically appraise alternative courses of action."

Answer: (Groupthink) Difficulty: (3)

Questions from the Appendix to Chapter 4

93. Break-even analysis is a financial analysis decision-making aid that enables a manager to determine how much profit a business venture will make.

Answer: (False) Difficulty: (2)

94. The break-even point is the point where the total revenue line crosses the total costs line.

Answer: (True) Difficulty: (1)

95. If a manager needs to design shipping schedules, assign personnel to various jobs, distribute merchandise from a number of warehouses to a number of customers, or route production to optimize the use of machinery, the manager might be wise to choose the mathematical method used to solve resource allocation problems called ________________.

Answer: (linear programming) Difficulty: (3)

Essay Questions

96. Describe the differences between programmed and non programmed decisions. Give an illustration of each.

Answer:

a. Programmed decisions are decisions that are repetitive and routine and that can be solved through mechanical procedures such as by applying rules. For example, to expedite the refund process, a department store may use the rule "If the customer returns the merchandise, you may give a refund if the tag is not removed, if the merchandise is not damaged, and if the purchase was made within the past two weeks."

These decisions are programmable, routine, generic, and computational. See Table 4.3 for additional information.

b. Non programmed decisions are unique and novel, and mechanical procedures are not available for making them. An example would be like managing the rescue work for a plane crash. Crises are not always required to be non programmable. See Table 4.3 for additional information.

Difficulty: (2)

97. List, briefly define, and briefly characterize or discuss the steps the "rational" manager would take in making a decision.

Answer:

See Figure 4.2 for a complete listing and diagram of the decision-making process.

a. Define the problem. Managerial decision making is usually sparked by the identification of a problem. The problem (what is at the root of the difficulty being encountered) should be separated from the symptoms of the problem.

b. Identify the criteria. In most decisions you'll want to achieve several objectives or satisfy several criteria. What is to accomplished by solving the problem? What aspects are important to the solution of the problem?

c. Weight the criteria. Some criteria might be more important than others (such as price). Rational decision makers will weight each of the decision criteria.

d. Develop alternatives. Whether we are choosing between alternative plans or job applicants, there are choices. The existence of some choice is a prerequisite to effective decision making. This may take time and creativity and judgment.

e. Analyze the alternatives. One expert says that this is the most difficult step. The decision maker must decide on how to evaluate the alternatives in a rational and honest way.

f. Make a choice, and then implement and evaluate the decision. Under perfect conditions, making a choice should be a straightforward matter of computing the pros and cons of each alternative relative to your decision criteria and choosing the one that maximizes your benefits.

Difficulty: (3)

98. In practice, many factors limit just how rational any decision maker can be. Several of these limits or barriers to rationality were discussed in your text. From the eight primary areas cited in your text, pick any <u>four</u> and discuss each as to primary limitations, primary characteristics, and give an illustration or example that will demonstrate your understanding of this critical area of understanding for managers.

Answer:

a. Individual differences. People's ability to absorb, analyze, and generally "process" information, while varying from person to person, is quite limited. An illustration is "information overload" that occurs when people try to remember and process constantly flowing information from a computer screen or projection screen. Systematic versus intuitive decision styles can also impact this area.

b. Decision-making shortcuts (heuristics). People take shortcuts by applying rules of thumb. This process undercuts what economists would desire from the rational decision maker. Managers make shortcut rules that generalize (such as "never allow a customer to spend up to the limit on their credit card"). Two forms of heuristics are possible: (a) similarity heuristics and (b) availability heuristics. Heuristics can allow for faster decisions, but they can also lead to errors.

c. Decision biases. There are two types of biases that impact decisions. First, there are person-specific biases and (second) cognitive biases. As indicated in the text, these biases allow for favoritism that is unwarranted. If one is not careful, a confirmation trap can occur. This means that people tend to seek confirmatory information for what they think is true and neglect to search for disconfirmatory evidence. This often occurs when a consumer wants to buy something and does not pay attention to evidence (such as price) that indicates that they should not make the purchase.

d. Framing the problem. The way the problem is presented (framing) can influence the rationality of the decision. The text used a lawsuit example to illustrate that the order of the wording alters the decision choice. This form of distortion is very common..

e. Escalation of commitment. Have you ever had to negotiate with someone only to find that both of your positions became more and more entrenched the more you negotiated? Sounds like dating—right? The solution is to have people make decisions in a step-by-step process rather than in total. Gas price wars often illustrate this problem.

f. Functional fixedness. This is the "inability to see that an object can have a function other than its stated or usual one." This decision-making barrier can limit a manager's ability to discover alternative solutions. Department managers tend to see a problem only from their own department's point of view (as an illustration).

g. Psychological set. This involves the tendency to focus on a rigid strategy or approach to certain types of problems. This can limit a manager's ability to seek or come up with alternative decisions. Perceptual distortion or thinking "within the box" is often an occurrence.

h. Organizational decision-making barriers. Sometimes it's the organization itself that inhibits the adequate evaluation of a problem and its possible solutions. For example, employees tend to be reluctant to bring bad news to their managers.

Difficulty: (3)

99. Define and characterize "groupthink."

Answer:

Groupthink has been defined as "a mode of thinking that people engage in when they are deeply involved in a cohesive group, in which the members' desire for unanimity overrides their personal motivation to realistically appraise alternative courses of action."

The classic groupthink situation was the Kennedy Administration's handling of the Bay of Pigs invasion.

You know you have a groupthink problem when such questions as "Don't ask, don't question," "You must conform," "We all agree," "We're on mission," and we are "Masters of the world" are common statements. In addition, Figure 4.9 presents 8 additional warning signs of groupthink. These are the illusion of invulnerability, collective rationalization, unquestioned morality, negative stereotyping, pressure to conform, self-censorship, illusion of unanimity, and mind guards.

Difficulty: (2)

Question from the Appendix to Chapter 4

100. Define linear programming and give an illustration of when it can be used.

Answer:

Linear programming is a mathematical method used to resolve allocation problems. These arise "whenever there are a number of activities to be performed, but also limitations on either the amount of resources or the way they can be spent."

Examples of usage include distribution of merchandise from a number of warehouses to a number of customers, assigning personnel to various jobs, designing shipping schedules, selecting the product mix in a factory to make the best use of machine and labor hours available while maximizing the firm's project, and routing production to optimize the use of machinery.

Difficulty: (2)

Chapter 5

PLANNING AND SETTING OBJECTIVES

INTRODUCTION

This chapter provides an overview of the planning process, the formulation of effective objectives, and the techniques for producing the premises on which plans are built.

LECTURE OUTLINE

Introduction

The Nature and Purpose of Planning
Why Plan?
Types of Plans
What Planning Accomplishes

The Management Planning Process
Steps In Planning
The Planning Hierarchy

How to Set Objectives
Why Set Objectives?
Types of Objectives
How to Set Motivational Goals
How to Express the Goal

Using the Management by Objectives Goal-Setting Technique

Techniques for Developing Planning Premises
Sales-Forecasting Techniques
Marketing Research
Competitive Intelligence

Planners in Action
Who Does the Planning?
What Planners Do
Planning in a Real Firm

Summary

Case: Planning at Cin-Made, Inc.
You Be the Consultant: KnitMedia Makes Plans to Expand

ANNOTATED OUTLINE

What's Ahead?

The Nature and Purpose of Planning

The situation faced by Jeff Bezos is an example of how the planning process increases in complexity as one moves up to the CEO function in an organization. Difficulties increase in depth and scope.

Why Plan? Plans are methods formulated beforehand for achieving a desired result. Plans specify goals and a course of action. Planning is the process of establishing objectives and courses of action prior to taking action. Goals or objectives are specific results to be achieved.

Notes:

Types of Plans. Plans have different formats, time horizons, and frequency. Descriptive plans might describe one's career plan, budgets are financial plans, and graphic plans use charts to provide data for analysis. Plans may be strategic from the top management level, usually for five to ten years, or tactical, usually for up to five years, or operational, for a day-to-day basis. Plans may be programs, usually designed for a one-time project, or standing as the need arises, e.g., policies, procedures, or rules.

Notes:

What Planning Accomplishes. Planning provides direction, reduces piecemeal decision making, reveals future opportunities and threats, and facilitates control of operations.

Notes:

The Management Planning Process

Steps in Planning. The steps in the planning include establishing objectives, analyzing the situation and creating planning premises, determining alternative courses of action, evaluating alternatives, and choosing/implementing the plan.

Notes:

The Planning Hierarchy. The hierarchy of plans includes the enterprise-wide plan and objectives and the chain of subsidiary unit plans and objectives that contribute to the achievement of enterprise-wide plans and objectives. A hierarchy of goals from the top level to the bottom is also included.

Notes:

How to Set Objectives

Why Set Objectives? Objectives provide a basis for appraisal of the performance of an organization and its employees. They provide targets toward which plans can be aimed and help provide motivation for individuals and teams.

Notes:

 Chapter 5, Planning and Setting Objectives 111

Types of Objectives. There are a number of types of objectives that can be set to gauge the performance of the organization and its employees. Objectives should be tailored to the specific organization.

Notes:

How to Set Motivational Goals. Goals are only useful if they provide motivation for employees to achieve them. Goal-setting studies report that goals should be specific, measurable, challenging, and encourage employee participation in the goal-setting process.

Notes:

How to Express the Goal. In expressing goals, planners should distinguish between an area to be measured, a yardstick and a goal. For example, an area could be sales, a yardstick could be sales revenue, and a standard/goal could be a dollar amount.

Notes:

Using the Management by Objectives Goal-Setting Technique

Management by Objectives (MBO). Management by objectives is a technique used by firms to assist in the process of setting organization-wide objectives and goals for subsidiary units and their employees. It is a technique in which a supervisor and a subordinate jointly set goals for the latter and periodically assess progress. The MBO process includes setting organizational goals, setting departmental goals, discussing departmental goals, setting individual goals, and the provision of feedback regarding the subordinate's performance. Goals should be reviewed and changed periodically to ensure they are current.

Notes:

Techniques for Developing Planning Premises

Sales Forecasting Techniques. Plans are based on planning premises or assumptions that people make about the future. A forecast is an estimate or calculation in advance, a prediction. Forecasting may be quantitative, e.g. by use of statistical methods, or qualitative, e.g., emphasizing human judgment. Quantitative methods include time-series analysis, where a set of observations are taken at specific times and used for prediction, and causal methods, where the relationship between two variables like sales and disposable income is examined. Qualitative methods include the jury of executive opinion where a jury of key executives forecast sales independent of each other. The president reconciles differences in forecasts. Another qualitative method is sales-force estimation where salespeople estimate their sales over the period in question.

Notes:

Marketing Research. Marketing research includes the procedures used to develop and analyze new customer-related information that helps managers make decisions. Marketing research experts use secondary data that consist of information that has already been collected or published and primary data that is collected to solve a current problem.

Notes:

Competitive Intelligence. Competitive intelligence is a systematic way to obtain and analyze public information about competitors. The use of this approach has tripled to about 10 percent of large companies since 1988. Greater use of this approach may increase the need for ethical, moral, and legal analysis in the planning process.

Notes:

Planners in Action

Who Does the Planning? Many organizations have moved from centralized to decentralized planning. The planners are generally not large groups of specialists at headquarters, but rather product and divisional managers often aided by small advisory groups at headquarters. These changes coincide with the general efforts to flatten organizations and decentralize decision making.

Notes:

What Planners Do. Planners act as information resources, conduct competitor and market research, develop forecasts, provide consulting services, create common reports and data for use across the organization, communicate values, communicate company-wide objectives, serve as planning-process guardians, develop planning methods, and lead planning teams.

Notes:

Planning in a Real Firm. The planning process at British American Tobacco Company (BAT) has a staff of four in its corporate planning office despite the large size of the organization with operations in 50 countries. The major role of the corporate planning department is to design and administer the planning process by monitoring business plans as they come in from the subsidiaries. The main role of the department is to formulate policies, objectives, and company wide strategies.

Notes:

Summary

Notes:

Case: Planning at Cin-Made, Inc.

1. Frey began by analyzing the market instead of establishing objectives. He seems to have determined objectives after performing the other steps in the process.

Chapter 5, Planning and Setting Objectives 113

2. Objectives could include:

a. To invest 20 percent more in R & D to implement the niche strategy in two years at a cost of $300,000.
b. To maintain high prices and maintain profitability levels during the next year at a cost of 10 percent fewer unit sales.
c. To hire 50 more salespeople in marketing to achieve custom products sales levels during the next year at a labor cost increase of $1 million.

3. Frey's planning method seems to be a hit and miss approach. He should follow the basic planning model in the chapter and begin by establishing objectives, performing a detailed analysis of the situation, and proceed to alternatives, choices, and implementation. His approach seems to have been successful in the short run but in the long run, it may lead to problems.

You Be the Consultant: KnitMedia Makes Plans To Expand

For the present time, budgets and financial projections for answering this assignment will have to be based on industry averages of operating and financial ratios for the entertainment and amusement industries, from information contained in Leo Troy, Almanac of Business and Industrial Financial Ratios (Englewood Cliffs, NJ, Prentice Hall, 1997). If and when KnitMedia goes public their own budgets will be listed on our Website. Useful average industry information for a small company in this industry will include the following:

Sales per employee: About $180,000
Operating costs as % of sales: Rent, 5%; Interest, 3:7%; taxes 4.4%; other, 50%
Ratio of debt to total assets: $1,000,000
Current assets/current liabilities:1.5
Profit Margin: 3%
Return on Assets: 5.5%

Also, assume for the sake of argument, the following for Knitmedia:

That Knitmedia has three main sources of revenues--The Knitting Factories, The Knit Works record label, and the Tours (including internet-based concerts)--and that each represents about 1/3 of the company's total sales; and,

- That KnitMedia assumes sales and profits will increase 30% over year in 1998, 1999, and 2000.
- That KnitMedia assumes they can open a total of five new Knitting Factories in 1998-2000.

1. The company has some very optimistic assumptions for sales and profit growth. An important factor in the ability to achieve the 30 percent per year increase would appear to be the ability of the firm to open five Knitting Factories in 1998-2000. The new operations will require a considerable amount of capital. Can the firm raise the money without going public? Assuming the firm would prefer to remain private, this would be an important question for consideration before planning to grow at such a rapid rate. Another important but highly uncertain factor is the ability to generate hit records during the period. Sales per employee for a small company in the entertainment and amusement industry are about $180,000. This seems quite high for Knit Media, unless they can generate a significant amount of revenue from the sales of a couple of hit records and successful concerts.

2. Plans developed by a firm are no better than the assumptions upon which they are based. Some of the assumptions the company should develop include (see page 164 of the textbook) : sales for each part

of the business; sales for each Knit Factory for the record label, and concert sales; competitors strategies, including possible expansion into Knit's business areas such as by large, well-funded companies such as Sony, or even Microsoft; ability to find and manage capable local management talent for clubs; and continued demand for Knit's unique type of music. Also (see textbook page 183): level of economic activity, political factors, such as problems with local zoning regulations; the impact of changing demographics in the general; and all of the technological changes taking place, such as more people using WebTV-type devices to use their TV's to access the internet and more companies like Viacom expanding their TV channel music offerings.

3. There are many assumptions Michael can use to develop better planning assumptions or premises. These include quantitative and qualitative sales forecasting techniques such as time-series analysis, causal methods, jury of executive opinion, and sales force estimation. The firm could also use marketing research tools and competitive intelligence. Good sources of marketing research information include internal company data on sales figures and invoices, observational research, survey/questionnaire research, and consulting experiments; government publications such as US Industrial Outlook; and periodicals such as Business Periodical Index, Standard and Poor's Industry Surveys, and Advertising Age; and commercial data, e.g., available from A.C, Nielson Company. The web site http://control1.bigcharts.com/report?r=sic&words=music&x=32&y=10 is an important source of marketing research information, lists of music companies, and their stock market performances. The site can also be used to link to other, similar financial web sites. Students can also check web sites for music companies including Universal, Sony, Warner and Viacom.

End of Chapter Questions for Chapter 5

Multiple Choice

1. ____________ are methods formulated beforehand for achieving a desired result.

a. Goals
b. Objectives
c. Plans
d. Policies
e. Rules

Answer: (c) Difficulty: (2)

2. All plans specify goals and ____________________

a. profits.
b. courses of action.
c. controls.
d. penalties for failure.
e. job functions.

Answer: (b) Difficulty: (2)

 Chapter 5, Planning and Setting Objectives

3. All plans specify goals. An example of a goal (as specified by a plan) would be (choose the best response) ____________________

a. "hire a new salesperson."
b. "develop an advertising theme that shows a caring attitude."
c. "meet people at the front door as they enter the store."
d. "boost sales by 10 percent."
e. "provide warranty information."

Answer: (d) Difficulty: (3)

4. All plans should specify (at a minimum) what you will do, how you will do it, and ______________

a. profits to be made.
b. employees to be used
c. by when you will get it done.
d. how much effort will be required.
e. how you will gain from the experience.

Answer: (c) Difficulty: (2)

5. ________ is (are) the "process of establishing objectives and courses of action, prior to taking action."

a. Goals
b. Objectives
c. Planning
d. Policies
e. Rules

Answer: (c) Difficulty: (2)

6. __________________ is often called the "first among equals" of the four management functions, because it establishes the goals which are (or should be) the basis of all these functions.

a. Organizing
b. Leading
c. Controlling
d. Planning
e. Human Resources

Answer: (d) Difficulty: (1)

7. Plans differ in format or in the way they are expressed. Which of the following plan types or formats "states in words what is to be achieved and how?"

a. Budgets
b. Graphic plans
c. Descriptive plans
d. A rule
e. An idea format

Answer: (c) Difficulty: (1)

8. All of the following are ways plans can be characterized <u>EXCEPT</u> (pick the least likely) ________

a. different profitability.
b. different formats.
c. different time horizons.
d. different frequency.
e. different levels of management authority needed.

Answer: (a) Difficulty: (3)

9. Plans differ in the spans of time they cover. Which of the following most accurately describes the form that top management uses to engage in long-term (five- to ten-year) plans? These plans normally specify what business or businesses the firm will be in.

a. Tactical plans
b. Strategic plans
c. Functional plans
d. Operational plans
e. Regional plans

Answer: (b) Difficulty: (2)

10. ________________ show how top management's plans are to be carried out at the department level (for instance, by the managers responsible for sales, finance, and manufacturing).

a. Tactical plans
b. Strategic plans
c. Future plans
d. Operational plans
e. Regional plans

Answer: (a) Difficulty: (2)

11. First-line managers focus on shorter-term plans. These might show, for instance, exactly which workers are to be assigned to which machines or exactly how many units will be produced on a given day. These plans are normally called ________________

a. Tactical plans.
b. Strategic plans.
c. Future plans.
d. Operational plans.
e. Regional plans.

Answer: (d) Difficulty: (3)

12. Planning involves different frequency of usage and application. Some plans are best characterized as being ________________ because they lay out in an orderly fashion all the steps in a major one-time project, each in its proper sequence.

a. programs
b. standing plans
c. procedures
d. rules
e. tactics

Answer: (a) Difficulty: (2)

13. In contrast to plans that are applied to a single use, ______________ are plans made to be used repeatedly, as the need arises.

a. programs
b. standing plans
c. strategic applications
d. operational applications
e. departmental applications

Answer: (b) Difficulty: (2) Page: 145

14. ________________ (as a form of standing plans) usually set broad guidelines for the enterprise. For example, at Saks Fifth Avenue the statement is made that "we sell only high-fashion apparel and top-of-the-line jewelry."

a. Policies
b. Procedures
c. Rules
d. Guidelines
e. Codes

Answer: (a) Difficulty: (2)

15. ________________ (as a form of standing plans) specify how to proceed if some specific situation arises. For example: "Before refunding the customer's purchase price, the salesperson should carefully inspect the garment and then obtain approval from the floor manager for the refund."

a. Policies
b. Procedures
c. Rules
d. Guidelines
e. Codes

Answer: (b) Difficulty: (3)

16. ________________ are usually considered to be highly specific guides to action. For example: "Under no condition will the purchase price be refunded after 30 days."

a. Policies
b. Procedures
c. Rules
d. Statements
e. Codes

Answer: (c) Difficulty: (2)

17. Implemented properly, planning can provide several advantages for the manager and the organization. All of the following are the best and most sure advantages of proper planning EXCEPT ________________

a. Planning insures profits.
b. Planning provides direction.
c. Planning reduces piecemeal decision making.
d. Planning reveals future opportunities and threats.
e. Planning facilitates control.

Answer: (a) Difficulty: (3)

18. Several year ago, R.R. Donnelly Company was mostly in the business of printing documents and other materials for clients such as investment bankers. Donnelly's planning led its managers to anticipate a change caused by telecommunications and the globalization of its customers. Donnelly then decided to reinvent itself. Which of the following advantages for planning is most likely the advantage that applies in the above example?

a. Planning insures profits.
b. Planning provides direction.
c. Planning reduces piecemeal decision making.
d. Planning reveals future competition.
e. Planning facilitates control.

Answer: (b) Difficulty: (3)

19. All of the following are part of the management-planning process EXCEPT ________________

a. Hire consultants to provide direction and objectivity.
b. Establish objectives.
c. Analyze situations/create planning premises.
d. Determine alternative courses of action.
e. Choose and implement the plan.

Answer: (a) Difficulty: (1)

Chapter 5, Planning and Setting Objectives

20. The first step in the management-planning process is to ____________________

a. hire consultants to provide direction and objectivity.
b. establish objectives.
c. analyze situations/create planning premises.
d. determine alternative courses of action.
e. choose and implement the plan.

Answer: (b) Difficulty: (1)

21. Al "Chainsaw" Dunlap recently took over Sunbeam Corporation (maker of small appliances). His first task was to develop a plan for turning the company around. He knew that within one year he had to make the company profitable, reduce expenses by at least 20 percent, and decide which products to keep and which to discard. Which of the following steps from the management-planning process most closely matches (is an example of) the above statement?

a. Hire consultants to provide direction and objectivity.
b. Establish objectives.
c. Analyze situations/create planning premises.
d. Determine alternative courses of action.
e. Choose and implement the plan.

Answer: (b) Difficulty: (2)

22. In practice, one result of the management-planning process is a ________________ which includes (1) the enterprise-wide plan and objectives and (2) the chain of subsidiary units' derivative plans and objectives, each of which contributes to achieving the enterprise-wide plans and objectives.

a. procedure
b. guideline
c. rule
d. document
e. hierarchy of plans

Answer: (e) Difficulty: (2)

23. A(n) ________________ summarizes the derivative targets to be achieved by each department if the long-term objective is to be met.

a. procedure
b. action plan
c. mandate
d. forecast
e. policy

Answer: (b) Difficulty: (1)

24. If there is one thing on which every manager can expect to appraised, it is the extent to which he or she achieves his or her unit's ________________ (Pick the best descriptive and most encompassing phrase.)

a. targets.
b. forecasts.
c. goals or objectives.
d. mandates.
e. expectations.

Answer: (c) Difficulty: (3)

25. ________________ are the targets toward which plans are aimed and the anchor-points around which the hierarchy of objectives is constructed. They also provide motivation.

a. Targets
b. Forecasts
c. Objectives
d. Mandates
e. Expectations

Answer: (c) Difficulty: (3)

26. The range of activities for which objectives may be set is virtually limitless. For example, in a classical analysis of planning objectives, Peter Drucker would say all of the following were examples of areas in which objectives should be set <u>EXCEPT</u> (pick the best and most broad from the available choices) ________________

a. unionization.
b. market standing.
c. innovation.
d. productivity.
e. profitability.

Answer: (a) Difficulty: (2)

27. Goals are only useful to the extent that employees are motivated to achieve them; managers can do several things to ensure that the goals they set do motivate employees. One way to gain insight into setting effective goals is to conduct organizational behavior research known as ________________ studies.

a. profitability
b. leadership potential
c. motivational
d. goal-application
e. goal-setting

Answer: (e) Difficulty: (3)

Chapter 5, Planning and Setting Objectives 121

28. Studies conducted by psychologists Edwin Locke and Gary Latham and their associates have provided valuable information on how managers set goals. From their studies several rules for proper goal setting have emerged. All of the following are rules that should be observed when setting goals EXCEPT ________________

a. assign goals that always lead to profits in the long-run.
b. assign specific goals.
c. assign measurable goals.
d. assign challenging but doable goals.
e. encourage participation where possible.

Answer: (a) Difficulty: (2)

29. Managers are faced with an interesting dilemma—should the managers assign their subordinates' goals, or should they permit their subordinates to participate in developing their own goals? Research evidence on this point has been mixed, however, all of the following are conclusions that have been reached with respect to participation in goal setting EXCEPT ________________

a. Employees who participate in setting their goals do in fact tend to perceive themselves as having had more impact on the setting of those goals than do employees who are simply assigned goals.
b. Participatively set goals tend to be higher (more difficult) than the goals the supervisor would normally have assigned.
c. Participatively set goals cause more misunderstandings of rules of behavior because of the fear of letting a superior know that misunderstanding was present.
d. Participatively set goals do not consistently result in higher performance than assigned goals.
e. Goals unilaterally assigned by managers can trigger employee resistance.

Answer: (c) Difficulty: (3)

30. Knowing how to express a goal is important. Three distinguishments may be made. They are the area to be measured, a(n) ________________, and a goal.

a. method
b. yardstick
c. evaluation
d. correlation
e. performance

Answer: (b) Difficulty: (2)

31. With respect to measuring and distinguishing goals, the terms "productivity," "number of rejects," "absenteeism," and "turnover" would all be examples of ________________

a. methods.
b. yardsticks.
c. evaluations.
d. correlations.
e. performances.

Answer: (b) Difficulty: (3)

32. Planning expert George Morrisey presents a four-point model for use in formulating objectives. According to Morrisey, a well-crafted goal should contain at least the following information: To (1) (action/accomplishment verb)...(2) (single measurable result) by (3) (target date/time span)....at (4) __________ (Pick the best answer.)

a. what profit.
b. which people.
c. cost in time and/or money.
d. resources used.
e. competition met.

Answer: (c) Difficulty: (3)

33. With respect to measuring and distinguishing goals, the terms or phrases "obtain all possible discounts," "zero complaints," "20% profit margins," and "90% of sales dollars collected per month" would be examples of ____________________

a. methods.
b. yardsticks.
c. evaluations.
d. correlations.
e. standards/goals.

Answer: (e) Difficulty: (3)

34. Using George Morrisey's criterion for (his four-point model) formulating a well-stated objective, which of the following statements best matches Morrisey's criterion?

a. The Acme project should be finished under budget.
b. Complete the Acme project at a cost not to exceed $50,000.
c. Release product A to manufacturing.
d. Reduce average turnaround time on service requests from eight to six hours by July 31 at an implementation cost of forty work-hours.
e. Try harder to improve quality control and cost containment in the Ft. Worth assembly plant under the direction of Mr. Jones.

Answer: (d) Difficulty: (3)

35. ________________ is a technique used by many firms to assist in the process of setting organization-wide objectives and goals for subsidiary units and their employees.

a. Management by objectives (MBO)
b. Profit projection
c. The Delphi Technique
d. Queuing Theory
e. Human Relations development

Answer: (a) Difficulty: (2)

Chapter 5, Planning and Setting Objectives 123

36. In common management terms and understanding, the abbreviation "MBO" stands for ________

a. measurable by organizations.
b. misled by objectives.
c. missions by organizations.
d. management by objectives.
e. management by organizations.

Answer: (d) Difficulty: (1)

37. ________________ is defined as a technique in which supervisor and subordinate jointly set goals for the latter and periodically assess progress toward those goals.
a. Management by organizations
b. Management by objectives
c. The Delphi Technique
d. Objectives simulation
e. Human Resource control

Answer: (b) Difficulty: (1)

38. One advantage of ________________ is that, if implemented properly, it can lead to specific, measurable, and participatively set challenging objectives.

a. management by organizations
b. management by objectives
c. the Delphi Technique
d. objectives simulation
e. human resource control

Answer: (b) Difficulty: (1)

39. The management by objectives (MBO) process consists of all the following steps EXCEPT ________

a. Review the employee's weaknesses based on past performance.
b. Set the organization's goals.
c. Set the department's goals.
d. Set individual goals.
e. Feedback.

Answer: (a) Difficulty: (1)

40. The management by objectives (MBO) process has all the steps listed below. Which of the steps is generally considered to be the first step in the process?

a. Feedback.
b. Set department's goals.
c. Set organization's goals.
d. Set individual goals.
e. Discuss department's goals.

Answer: (c) Difficulty: (2)

41. Managers can do several things to make an MBO program more successful. All of the following would be useful EXCEPT (pick the least likely) ____________________

a. State the goals in measurable terms.
b. Be specific with respect to goals.
c. Apply pressure to make sure that deviation will not occur.
d. Make sure each person's goals are challenging by attainable.
e. Goals should be reviewed and updated continually.

Answer: (c) Difficulty: (2)

42. Which of the following statements might be made by Peter Drucker (with respect to MBO)? To answer, consider what Peter Drucker feels about MBO and its place in the modern organization.

a. "The goals of the individual are more important than the goals of the organization."
b. "The goals of each manager's job must be defined by the contribution he or she has to make to the well-being of each employee and his/her family."
c. "The MBO process simply does not work in the American firm."
d. "The MBO process simply does not work in the average Japanese firm."
e. "The job objectives of district sales managers should be defined by the contribution they and their district sales forces have to make to the sales department."

Answer: (e) Difficulty: (3)

43. Managers use several techniques to produce the premises on which they build their plans. The most popular techniques are forecasting, marketing research, and ______________________.

a. soothsaying.
b. competitive infiltration.
c information system management.
d. secondary research.
e. competitive intelligence.

Answer: (e) Difficulty: (3)

44. Most plans are based on ____________________, which (is) are the assumptions we make about the future.

a. soothsaying
b. secondary research
c. information system management
d. planning premises
e. competitive infiltration

Answer: (d) Difficulty: (2)

Chapter 5, Planning and Setting Objectives 125

45. A(n) _____________ means to estimate or calculate in advance or to predict.

a. forecast
b. plan
c. operation chart
d. correlation
e. survey

Answer: (a) Difficulty: (1)

46. According to one expert, a(n) _______________ is "a service whose purpose is to offer the best available basis for management expectations of the future and to help management understand the implications for alternative courses of action."

a. forecast
b. plan
c. operation chart
d. correlation
e. survey

Answer: (a) Difficulty: (1)

47. All of the following are popular sales forecasting methods <u>EXCEPT</u> _____________

a. time-series analysis.
b. causal models.
c. performance estimation juries.
d. jury of executive opinion.
e. sales force estimation.

Answer: (c) Difficulty: (2)

48. _________________ forecasting methods use statistical methods to examine data and find underlying patterns and relationships.

a. Qualitative
b. Quantitative
c. Projective
d. Illustrative
e. Synthesis

Answer: (b) Difficulty: (2)

49. _____________________ methods emphasize human judgment.

a. Qualitative
b. Quantitative
c. Projective
d. Illustrative
e. Synthesis

Answer: (a) Difficulty: (2)

50. Which of the following is an example of a quantitative forecasting technique?

a. Delphi method
b. Causal models
c. Performance estimation juries
d. Jury of executive opinion
e. Sales force estimation

Answer: (b) Difficulty: (3)

51. Which of the following is an example of a qualitative forecasting technique?

a. Delphi method
b. Time-series method
c. Casual models
d. Jury of executive opinion
e. Rate of advertising decay

Answer: (d) Difficulty: (3)

52. A ____________ analysis is a set of observations taken at specific times, usually at equal intervals.

a. Delphi method
b. time-series
c. causal model
d. jury of executive opinion
e. rate of advertising decay

Answer: (b) Difficulty: (2)

53. Examples of _________________ are the yearly or monthly gross domestic product (GDP) of the United States over several years, a department store's total monthly sales receipts, and the daily closing prices of a share of stock.

a. the Delphi method
b. time-series
c. a causal model
d. the jury of executive forecasting
e. rate of pricing decay

Answer: (b) Difficulty: (2)

54. The basic purpose of all time-series forecasting methods is to _________________

a. track sales.
b. remove irregular and seasonal patterns so that management can identify fundamental trends.
c. know the sequence of events taking place in the management cycle.
d. determine negative coefficients and "weed" them out of analysis.
e. determine negative coefficients and be able to convert them into position coefficients.

Answer: (b) Difficulty: (3)

55. Several methods are used to isolate a time-series' underlying trend. ___________ methods average the data in some way to remove seasonal and random variations.

a. Simple moving average
b. Causal
c. Exponential
d. Estimation charts
e. Smoothing

Answer: (e) Difficulty: (2)

56. _______________ methods develop a projection based on the mathematical relationship between a company factor and those variables that management believes influence or explain the company factor.

a. Simple moving average
b. Causal
c. Exponential
d. Estimation charts
e. Smoothing

Answer: (b) Difficulty: (2)

57. The basic premise of _______________ methods is that a particular factor (such as your company's sales of television sets) is directly influenced by some other, more predictable factor or factors (such as the number of people unemployed in your state or the level of disposable income in the United States).

a. simple moving average
b. causal
c. exponential
d. estimation charts
e. smoothing

Answer: (b) Difficulty: (2)

58. All of the following are often cited as disadvantages of using or relying on quantitative methods of forecasting such as time series or causal methods EXCEPT (pick the least likely disadvantage) ____________________

a. They are useless when data are scarce.
b. They are useless if no history for the data or product is to be found.
c. They assume historical trends will continue.
d. They are extremely costly to undertake.
e. They tend to disregard unforeseeable, unexpected occurrences.

Answer: (d) Difficulty: (3)

59. The ___________________ method involves asking executives to make a forecast based on data that concerns economic levels and changes that might occur in the economy or company (such as with products or services).

a. Delphi
b. time-series
c. causal model
d. jury of executive forecasting
e. rate of pricing decay

Answer: (d) Difficulty: (1)

60. The ___________________ method gathers opinions of the sales force regarding what they think some factor (say, sales) will be in the forthcoming period. The separate estimates are then combined into a single sales forecast for the firm.

a. Delphi
b. time-series
c. causal model
d. jury of executive forecasting
e. sales force estimation

Answer: (e) Difficulty: (2)

61. An often cited disadvantage of the sales force estimation method is that a salesperson may (pick the most likely disadvantage) ___________________

a. cheat.
b. be unreliable.
c. not be capable of forecasting.
d. underestimate sales if they know that the estimate will become their quota.
e. not be familiar with the territory and unable to accurately forecast (as is the case with the new salesperson).

Answer: (d) Difficulty: (2)

62. One way to manage change under conditions of uncertainty is to make projections based on _______________ planning (defined as hypothetical sequences of events constructed for the purpose of focusing attention on causal processes and decision points).

a. perception
b. role-playing
c. scenario
d. extrapolation
e. fantasy

Answer: (c) Difficulty: (2)

 Chapter 5, Planning and Setting Objectives

63. Some companies such as Shell Oil must make major investments, although the future today may be very unpredictable. By creating ____________ such as the "sustainable world" and "global mercantilism," the company can prepare itself for any shift in the world environment and level of competition.

a. perceptions
b. role-playing events
c. scenarios
d. extrapolations
e. fantasies

Answer: (c) Difficulty: (2)

64. ________________ refers to the procedures used to develop and analyze new customer-related information that helps managers make decisions.

a. Market segmentation
b. Information gathering
c. Competitive intelligence
d. Data preview
e. Marketing research

Answer: (e) Difficulty: (2)

65. ________________ data refer to information specifically collected to solve a current problem.

a. Secondary
b. Primary
c. Product
d. Executive
e. Managerial

Answer: (b) Difficulty: (1)

66. Examples of ________________ data sources include mail and personal surveys, in-depth and focus-group interviews, and personal observations.

a. secondary
b. primary
c. product
d. executive
e. managerial

Answer: (b) Difficulty: (2)

67. __________ is a systematic way to obtain and analyze public information about competitors.

a. Market segmentation
b. Information gathering
c. Competitive intelligence
d. Data preview
e. Marketing research

Answer: (c) Difficulty: (1)

68. All of the following would be illustrations of ethical and legal concerns that management could pursue with competitive intelligence EXCEPT (pick the least likely) ________________

a. information about competition.
b. information about business relationships and transactions.
c. information about executives' private lives and moral behavior.
d. information about entry into new markets.
e. information about sales opportunities.

Answer: (c) Difficulty: (1)

69. All of the following are listed as basic planning-related activities at General Electric EXCEPT (pick the item that would not be characterized as a typical primary planning activity) ______________

a. conduct competitor and marketing research.
b. develop forecasts.
c. communicate values.
d. develop planning methods.
e. develop new ways to lobby Washington legislators

Answer: (e) Difficulty: (2)

True/False Short Answer

70. All plans specify goals (such as "boost sales by 10 percent") and amount of profits to be made (such as "all sales must yield a 10 percent profit").

Answer: (False) Difficulty: (3)

71. ______________ is the process of establishing objectives and courses of action.

Answer: (Planning) Difficulty: (3)

72. Planning is the process of establishing objectives and courses of action.

Answer: (True) Difficulty: (1)

73. Once planning has taken place, goals or objectives are specific results you want to achieve.

Answer: (True) Difficulty: (1)

 Chapter 5, Planning and Setting Objectives

74. Planning is often called "the first and most important step" of the four management functions.

Answer: (False) Difficulty: (3)

75. As long as planning is sound, the other management functions tend to take care of themselves.

Answer: (False) Difficulty: (2)

76. ______________ plans (such as a career plan) state in words what is to be achieved and how.

Answer: (Descriptive) Difficulty: (3)

77. ________________ plans show in charts what is to be achieved and how.

Answer: (Graphic) Difficulty: (3)

78. Middle managers typically focus on constructing strategic plans because of the popular movements toward decentralization.

Answer: (False) Difficulty: (2)

79. Operational plans are often described as being the detailed day-to-day plans necessary to make a company work.

Answer: (True) Difficulty: (1)

80. Rules set broad guidelines for an organization.

Answer: (False) Difficulty: (2)

81. One of the primary advantages of planning is that it facilitates control.

Answer: (True) Difficulty: (1)

82. Normally, the first step in the management-planning process is to review budget performance so one knows how many resources are available for planning.

Answer: (False) Difficulty: (3)

83. Normally, the first step in the management-planning process is to ______________.

Answer: (establish objectives) Difficulty: (2)

84. Research has shown that when performance is inadequate, it is often because the person or team is loafing.

Answer: (False) Difficulty: (2)

85. An example of a yardstick that can be used to express information about a goal would be employee behavior.

Answer: (False) Difficulty: (3)

86. According to George Morrisey's four-point model for expressing well-stated objectives, the following example would contain all the vital elements of the well-stated objective: "To complete the Acme project at a cost not to exceed $50,000 and five hundred work-hours."

Answer: (False—lacks a target date/time span) Difficulty: (2)

87. The term "MBO" almost always refers to a program done within a single department to establish objectives for that department's employees.

Answer: (False) Difficulty: (3)

88. The quantitative technique of forecasting that uses smoothing is called ________________ analysis.

Answer: (time-series) Difficulty: (3)

Essay

89. Define and discuss the different forms or types of planning and plans.

Answer:

a Planning is the process of establishing objectives and courses of action, prior to taking action. Goals or objectives are the specific results you want to achieve. Planning is an activity in which we all engage every day, often without giving it much thought.

b. There are many types of plans. They can be classified by three means: (1) different formats, (2) different time horizons, and (3) different frequency.

c. Different formats: Plans differ in format or in the way they are expressed. They can be descriptive (stating in words what is to be achieved—like a career plan), they can be stated in financial terms (like a budget), or they can be graphic (shown in charts as to what is to be achieved and how).

d. Different time horizons: Plans differ in the spans of time they cover. They can be top-management 5- to 10-year strategic plans (which define what business the firm will be in and what steps it must take to get there), or middle management shorter-term tactical plans (up to five-year durations) that show how management's top plans are to be carried out at the department level, or lower-level operational plans that detail day-to-day activities.

e. Different frequency: Plans can be programs (laying out steps for a one-time major project), standing plans (made to be used frequently), policies (setting forth broad guidelines for the enterprise), procedures (how to proceed if a special situation arises), or rules (a highly specific guide to action).

Difficulty: (2)

90. List and briefly identify the advantages that planning accomplishes.

Answer:

a. Planning provides direction. Planning provides direction and a sense of purpose for the enterprise.

Chapter 5, Planning and Setting Objectives

b. Planning reduces piecemeal decision making. A plan also provides a unifying framework against which decisions can be assessed. A sequence is provided.

c. Planning reveals future opportunities and threats. Opportunities and risks can be identified. Long-term risks can be minimized with proper planning.

d. Planning facilitates control. Control means ensuring that activities conform to plans. It is a three-step process whereby standards are set, performance is measured against these standards, and deviations are identified and corrected.

Difficulty: (1)

91. List and briefly summarize the management-planning process.

Answer:

a. Establish objectives. In a business, the first step in planning is to establish objectives for the entire enterprise and then, through the process of planning, to establish them in turn for subordinate units, each of which will develop their own plans.

b. Analyze the situation and create planning premises. This step develops forecasts and planning premises on which the plan will be built. Look at the environment and processes within the firm.

c. Determine alternative courses of action. There is obviously more than one way to proceed. This step devises several courses of action that might be possible.

d. Evaluate alternatives. Each alternative should be evaluated relative to the manager's objectives.

e. Choose and implement the plan. It is time to choose and act in this step. Feedback can be useful to carefully monitor the progress of the plan so planning will be easier and more efficient the next time it becomes necessary.

Difficulty: (2)

92. Knowing how to express goals and objectives is important. It is important to distinguish between an area to be measured, a yardstick, and a goal. Using George Morrisey's four-point model for use in formulating objectives, create an example that demonstrates the use of each of his four points. Then, indicate the area to be measured, a yardstick for doing so, and a goal. Be precise in your comments and directly tie your example (for example, 1, 2, 3, 4) to the Morrisey format.

Answer:

The instructor should review Table 5.3 and Table 5.4 for illustrations of areas, yardsticks, and standards/goals and demonstrations of the four-point model.

As an example: To (1, 2) complete the Acme project by (3) December 31 at a (4) cost not to exceed $50,000 and five hundred work-hours.

Area: Expenses Yardstick: materials Standard/goal: $50,000 and five hundred work-hours.

Difficulty: (3)

93. Define management by objectives (MBO) and list and briefly discuss how the steps that MBO generally consists of apply to this method.

Answer:

Management by objectives (MBO) may be defined as being a technique used by many firms to assist in the process of setting organization-wide objectives and goals for subsidiary units and their employees. It is further explained as a technique in which supervisor and subordinate jointly set goals for the latter and periodically assess progress toward those goals.

The MBO process generally consists of five steps:

a. Set the organization's goals. Here top management sets strategic goals for the company.

b. Set departments' goals. Department heads and their superiors jointly set supporting goals for their departments.

c. Discuss departments' goals. Department heads present departments' goals and ask all subordinates to develop their own individual goals.

d. Set individual goals. Goals are set for each subordinate, and a timetable is assigned for accomplishing those goals.

e. Feedback. The supervisor and subordinate meet periodically to review the subordinate's performance and to monitor and analyze progress toward his or her goals.

Difficulty: (3)

Chapter 6
STRATEGIC MANAGEMENT

INTRODUCTION

This chapter includes analysis of the strategic management process, situational analysis, types of strategies, and strategy implementation. Top management must be successful in formulating and implementing strategy to accomplish the mission of the organization and achieve organizational goals.

LECTURE OUTLINE

What's Ahead?

The Strategic Management Process
- Step 1: Define the Business and Develop a Mission Statement
- Step 2: Translate the Mission into Strategic Objectives
- Step 3: Formulate a Strategy to Achieve the Strategic Objectives
- Step 4: Implement the Strategy
- Step 5: Evaluate Performance and Correct as Required
- The Strategic Planning Process

Situation Analysis
- SWOT Analysis
- The Environmental Scan
- Benchmarking

Types of Strategies
- Corporate Level Strategies
- Competitive Strategies
- Competitive Analysis: The Five Forces Model
- Strategy and the CEO
- Functional Strategies

Implementing the Strategy
- Achieving "Strategic Fit"
- Strategy as Stretch and Leverage
- The Strategic Role of Core Competencies
- How Entrepreneurs Craft Strategies

Summary

Case: Harrison Products and the Question of Alliances
You Be the Consultant: KnitMedia's Strategies for Growth and Expansion

ANNOTATED OUTLINE

What's Ahead?

How can Sun Microsystems, Inc. maintain its strategic advantage? Part of the answer is internal, part is in forming strategic alliances with other computer firms. Sun needs a coherent strategy. For that matter, every firm should engage in the strategic management activities covered in this chapter.

The Strategic Management Process

Strategic management is the process of identifying and pursuing the organization's mission by aligning the organization's internal capabilities with the external demands of its environment.

Step 1: Define the Business and Develop a Mission Statement. The first step is to define the business and develop a mission statement. Top management should develop a vision statement, or a general statement of the intended direction of the enterprise that evokes emotional feelings in organization members. A mission statement that further defines the company's purpose and operationalizes the top management vision should also be developed

Notes:

Step 2: Translate the Mission into Strategic Objectives. The second step is to translate the vision and mission statements of top management into operational strategic objectives.

Notes:

Step 3: Formulate a Strategy to Achieve the Strategic Objectives. A strategy is a plan of action that will take the organization from where they are now to where they want to be. This plan must consider not only the organization's internal strengths and weaknesses but the opportunities and threats in the external environment.

Notes:

Step 4: Implement the Strategy. Put the plan into action. This will require use of the management functions of planning, organizing, leading, and controlling.

Notes:

Step 5: Evaluate Performance and Correct as Required. The final step includes strategic control, assessing progress toward strategic objectives and making corrections as required.

Notes:

The Strategic Planning Process. Strategic planning is a part of the overall planning process. It involves the process of identifying the business of the firm currently and where it intends to go in the future. The strategic planning process is a series of steps including definition of the business and a mission statement, establishment of strategic objectives, formulation of strategies to achieve strategic objectives, implementation of strategies, and evaluation to correct as required.

Notes:

Situational Analysis

Situation analysis assesses the internal and external forces that will affect most significantly the strategic options and opportunities for the organization. Situation analysis aids include SWOT analysis, environmental scanning, and benchmarking.

Notes:

SWOT Analysis. SWOT analysis is a technique for summarizing strengths, weaknesses, opportunities, and threats facing an organization. Formulating a strategic plan is largely a process of identifying strategies that will balance internal strengths and weaknesses with external opportunities and threats.

Notes:

The Environmental Scan. Environmental scanning is a systematic method of obtaining and compiling information about environmental forces that might be relevant to the strategic planners in the company. Areas normally scanned include economics, politics, social trends and demographics, technology, competition and geography.

Notes:

Benchmarking. Benchmarking is a process where a company learns how to improve its outcomes by carefully analyzing the practices of other companies who excel in the outcome desired. Companies using benchmarking should focus on a specific problem and define it carefully, use employees who will implement the changes to study how the benchmark organization operates, be willing to share information with the benchmark organization, avoid sensitive issues such as pricing, and maintain the confidentiality of information received.

Notes:

Types of Strategies

Corporate-Level Strategies. Corporate-level strategy identifies the portfolio of businesses that will comprise the corporation and how these businesses will relate to each other.

Notes:

Concentration. In a concentration corporate strategy, the company focuses on one product or product line, usually in one market. With this strategy, the company can focus on the business that it knows well and attempt to be the best in that arena. The risk of concentrating on one business can be disastrous if that business declines. Four strategies can contribute to growth via concentration: market penetration, geographic expansion, product development, or horizontal integration.

Notes:

Vertical Integration. Vertical integration means owning or controlling the inputs to the firm's processes and/or the channels through which the firm's products or services are distributed. Owning or controlling the inputs is known as backward vertical integration and owning or controlling the distribution channels is known as forward vertical integration.

Notes:

Diversification. Diversification is a strategy of expanding into related or unrelated products or market segments. This strategy avoids the risk of concentrating in only one line of business. Related diversification means diversifying into other industries that fit the lines of business of the firm. Conglomerate diversification means diversifying into products or markets unrelated to the present business of the firm.

Notes:

Status Quo Strategies. An organization implements status quo strategies when it is satisfied with its present rate of growth and intends to continue focusing on its present products and markets.

Notes:

Investment Reduction Strategies. These strategies are implemented when a firm has over expanded or has some form of financial emergency. Corrective actions are taken to reduce company investment in one or more of its lines of business. Retrenchment is the reduction of activity or operations. Divestment means selling or liquidating individual businesses.

Notes:

Strategic Alliances and Joint Ventures. Firms can grow by forming a partnership with another company through strategic alliances and joint ventures. Organizations can benefit through utilizing each other's strengths in a complementary fashion.

Notes:

The Virtual Corporation. A virtual corporation is a temporary network of independent companies--suppliers, customers, rivals linked by information technology to share skills, costs, and access to each other's markets. These organizations do not have a headquarters and chain of command like traditional corporations. Successful virtual corporations are built on a sense of trust and co-destiny.

Notes:

Corporate Strategy Portfolio Analysis Tools. Portfolio analysis tools have been developed, including the BCG (Boston Consulting Group) Matrix and the GE Business Screen. The BCG Matrix identifies the relative attractiveness of each business in the firm. It compares business growth rate and relative competitive position for each business. Decisions are made as to whether each business is a star, question mark, cash cow, or dog. A star is a high-growth business with high market share. Question marks are in high-growth industries but have low market share. A cash cow has high market share in a low-growth industry. Dogs are low-market-share businesses in low-growth, unattractive industries. The GE Business Screen has a nine-cell matrix where firms are plotted in a cell based upon industry attractiveness and business unit position.

Notes:

Competitive Strategies. Competitive strategies identify how to build and strengthen the long-term competitive position in the marketplace. A competitive strategy is a plan to establish a profitable and sustainable competitive position against the forces that determine industry competition. Each business in the firm needs a competitive strategy.

Notes:

Cost Leader. A business that pursues a cost leadership competitive strategy is planning to become the low-cost leader in an industry.

Notes:

Differentiator. As a differentiator, a firm seeks to be unique in its industry along a dimension valued by buyers, e.g., Volvo by emphasizing safety in its cars.

Notes:

Focuser. A focuser selects a market segment and builds its competitive strategy on serving the customers in its market niche better or more cheaply than its competitors.

Notes:

Competitive Analysis: The Five Forces Model. The competitive intensity reflects five competitive forces in the Five Forces Model. These include threat of entry, the intensity of rivalry among existing competitors, pressure from substitute products, the bargaining power of buyers, and the bargaining power of suppliers. The Five Forces Model can be used to choose competitive strategy options.

Notes:

Strategy and the CEO. This section discusses how a CEO's values and personality have a prevailing effect on the strategy he or she chooses. Miles and Snow found four characteristic types of strategy makers, including defenders, prospectors, analyzers, and reactors. Defenders are pursuing concentrator/specialization strategies to become the best at producing and marketing a specialized product or service. Prospectors run firms that are continually searching for new market opportunities and experimenting with new products or services. Analyzers are more cautious in strategy formulation and try to combine the efficiency advantage of defenders with the innovation and new-product development advantage of prospectors. Reactors are the most cautious in strategy formulations, and they tend to pursue status quo strategies.

Notes:

__Functional Strategies__. Functional strategies identify the basic courses of action that each of the functional departments will pursue to contribute to attaining the goals of the business. They flow from the competitive strategies and strategic goals of the overall organization.

Notes:

Implementing the Strategy

__Achieving "Strategic Fit"__. All of the firm's activities must be tailored to fit the strategy of the organization. For example, if the firm plans to be a low-cost producer, its activities must focus on efficiency and cost controls.

Notes:

__Strategy as Stretch and Leverage__. Although firms must effect a fit between resources and opportunities, preoccupation with fit can unnecessarily limit the growth of a company. Some experts believe that stretch should supplement fit. These researchers believe that leveraging resources can be more important than just fitting the strategic plan to the current resources of the firm.

Notes:

__The Strategic Role of Core Competencies__. This summarizes the research of Prahalad and Hamel who believe core competencies should be leveraged. Core competencies are the collective learning in the organization, especially knowing how to coordinate diverse production skills and integrate multiple streams of technologies.

Notes:

__How Entrepreneurs Craft Strategies__. Entrepreneurs use three general guidelines in formulating their strategies. These include quickly screening out losers by applying judgement and reflection, minimizing resources devoted to researching ideas, and being ready to change without waiting for all of the answers.

Notes:

Summary

Case: Harrison Products and the Question of Alliances

1. The suitors are recommending a strategic alliance or joint venture. This could solve the marketing problems of Harrison's firm and provide a more competitive organization for marketing efforts.

2. The threat of entry should not be too severe because of the high technology involved. Other firms are not going to be able to easily develop competitive products. The bargaining power of suppliers or buyers should not be a significant problem. The threat of substitute products does not seem to be a significant concern. The intensity of rivalry among existing competitors is quite high, and this is a major concern to Harrison because of the lack of a large, global sales force to compete with the larger firms in the market. Perhaps the firm could implement a diversification strategy and attempt to make some of the products that its competitors proposed in the strategic alliance, thereby marketing the whole system. It could also maintain its current specialized product line of filters but limit its marketing efforts to one geographic area to use it's sales force more effectively. It could also try to develop its own sales force rather than depending upon independent sales representatives.

3. An environmental scan would need to include the markets within which the firm plans to compete. If it is to compete globally, it must scan the economic, political, social trends/demographics, technology, and competition on a global basis. If it plans to concentrate on a particular country, it can narrow the focus of its scan to that environment. It needs to focus forecasting efforts, marketing research, and competitive intelligence-gathering to the market in which it plans to compete. The results of a global environmental scan might force the firm to operate in a more specific geographic market where its sales force is more likely to be successful. To expand on a global basis, if it does not enter a strategic alliance, it may have to hire and develop its own global sales force.

You be the Consultant: KnitMedia's Strategies for Growth and Expansion

1. In terms of generic corporate strategies, KnitMedia is pursuing Geographic expansion (for instance with its clubs); Vertical integration, (for instance by opening clubs as venues for artists it signs for its record labels); and (related) diversification, (by embarking on several related businesses such as tours and festival promotion). It would be useful for students to compare Knit's range of businesses with those of competitors; on balance, particularly given the relatively narrow music genre that Knit limits itself to the strategy of having an interlocking group of companies all pursuing different ways of utilizing Knit's alternative -jazz brand image is a sound strategy. One potential, but unlikely downside: Knit's brand is tied closely to alternative jazz: should the market for this type of music start to shrink, having all it's businesses tied to it might hit Knit relatively hard.

2. It already seems to be achieving the vertical integration it wants, for instance by showcasing its record-label artists in its clubs. By raising more money through a stock offering, Knit could sign even more great artists, showcase them in more clubs, and afford to sponsor more tours. One possible problem with showcasing its record-label artists in the clubs is the possibility that they may be pirated by competitors. On the other hand, exposure in the Clubs may be an important factor in attracting new talent to Knit Media.

End of Chapter Questions for Chapter 6

Multiple Choice

1. The strategic management process has received attention from management writers for many years. Management expert __________________ has said that the primary task of top management is "thinking through the mission of the business, that is, of asking the question 'What is our business, and what should it be?'"

a. Milton Friedman
b. John Maynard Keynes
c. Frederick W. Taylor
d. Peter Drucker
e. Tom Peters

Answer: (d) Difficulty: (2)

2. According to Peter Drucker, the process of setting objectives, the development of strategies and plans, and the making of today's decisions for tomorrow's results is possible when management asks and answers the question:

a. "What is our stage in the product life cycle?"
b. "Who is in charge, and how competent are our leaders?"
c. "How will information technology impact our organization?"
d. "How will we compete with global competitors?"
e. "What is our business, and what should it be?"

Answer: (e) Difficulty: (3)

3. ________________ is the process of identifying and pursuing the organization's mission by aligning the organization's internal capabilities with the external demands of its environment.

a. Management
b. Functional management
c. Strategic management
d. Goal formulation
e. Organizational behavior

Answer: (c) Difficulty: (2)

4. All of the following tasks are part of the strategic management process EXCEPT ______________

a. Defining the business, developing a vision, and mission.
b. Using quantitative models to solve functional problems in a qualitative environment.
c. Crafting a strategy to achieve the objectives.
d. Implementing and executing the strategy.
e. Initiating corrective adjustments.

Answer: (b) Difficulty: (2)

5. The first step in the strategic management process is to ____________________

a. craft a strategy to achieve the objectives of the firm.
b. develop MBO.
c. use quantitative forecasting models to chart the future of the firm.
d. define the business and develop a mission statement.
e. develop an organizational hierarchy.

Answer: (d) Difficulty: (2)

6. If two companies make the basic statement, "We make cars," then we know that they (pick the best answer of those available) _______________

a. are in the same business.
b. have the same mission statements.
c. have the same objectives.
d. have the same strategies.
e. view profits in the same way.

Answer: (a) Difficulty: (1)

7. K-Mart and Wal-Mart approach their businesses in two different ways. Wal-Mart has distinguished itself by concentrating its stores in small southern towns and by building a state-of-the-art satellite-based distribution system. K-Mart, on the other hand, opened stores throughout the country (where competition was more intensified). The above is an illustration of (pick the best answer of those available) ____________________

a. small versus large strategic application.
b. same business, different missions.
c. price competition versus distribution competition.
d. a one-price policy versus a discounter strategy.
e. similarity but regionalization.

Answer: (b) Difficulty: (3)

8. Which of the following would NOT be an illustration of a good mission statement?

a. To provide a high reliability, error-free method for moving people and products up, down and sideways within a building (Apex Elevator)
b. To provide information service in local-exchange and exchange-access markets within its franchise area, as well as cellular phone and paging services (United Telephone Corporation of Dade)
c. Knowledge is good (Faber College)
d. To provide people with longer lives and higher-quality lives by applying research efforts to develop new or improved drugs and health care products (Josephson Drug Company, Inc.)
e. To transform how educators work by providing innovative and easy-to-use multi-media-based computer systems (Gray Computer, Inc.)

Answer: (c) Difficulty: (1)

9. Which of the following would NOT be an illustration of a good mission statement?

a. To provide a high-reliability, error-free method for moving people and products up, down and sideways within a building (Apex Elevator)
b. To provide information service in local-exchange and exchange-access markets within its franchise area, as well as cellular phone and paging services (United Telephone Corporation of Dade)
c. Progress through better electricity (General Electric)
d. To provide people with longer lives and higher-quality lives by applying research efforts to develop new or improved drugs and health-care products (Josephson Drug Company, Inc.)
e. To transform how educators work by providing innovative and easy-to-use multimedia-based computer systems (Gray Computer, Inc.)

Answer: (c) Difficulty: (2)

10. Management must consider several issues when defining their business. In turn, by so doing, they answer the question, "What business should we be in?" The answer to this question usually results in both a _______________ statement and a mission statement.

a. profit
b. functional
c. directional
d. leadership
e. vision

Answer: (e) Difficulty: (2)

11. Management must consider several issues when defining their business. In turn, by so doing, they answer the question, "What business should we be in?" The answer to this question usually results in both a _______________ and a _____________ statement.

a. profit; mission
b. functional; profit
c. directional; challenge
d. leadership; vision
e. vision; mission

Answer: (e) Difficulty: (3)

12. The organization's _______________ statement is a "general statement of its intended direction that evokes emotional feelings in organization members."

a. vision
b. mission
c. strategy
d. belief
e. challenge

Answer: (a) Difficulty: (2)

13. According to Warren Bennis and Bert Manus, the critical point in the early stages of the thought process about directions for the organization is that "a ___________ articulates a view of a realistic, credible, attractive future for the organization, a condition that is better in some important ways than what now exists."

a. vision
b. mission
c. strategy
d. belief
e. dyad challenge

Answer: (a) Difficulty: (2)

14. Consider the following examples of industry innovators and their statements about their industries and companies: (1) Dr. Edwin Land (who invented the Polaroid camera) said a company could be built of providing instant photographs in self-contained cameras; (2) Rupert Murdock thought of an integrated global news- gathering, entertainment, and multimedia firm; and (3) Bill Gates perceived of a software company serving the needs of the then-fledgling microcomputer industry. All of these developers had sound __________ statements.

a. belief
b. global perspectives
c. vision
d. challenge/response
e. profit/loss

Answer: (c) Difficulty: (2)

15. A firm's _______________ statement further defines the company's purpose and operationalizes the top manager's vision.

a. belief
b. mission
c. leadership
d. profit
e. functional

Answer: (b) Difficulty: (1)

16. A _____________ statement "broadly outlines the organization's future course and serves to communicate 'who we are, what we do, and where we're headed.' "

a. belief
b. mission
c. leadership
d. profit
e. functional

Answer: (b) Difficulty: (2)

17. Consider the following statements about Citicorp: (1) building shareholder value through sustained growth in earnings per share; (2) continuing commitment to building customer-oriented business worldwide; (3) maintaining superior rates of return; and (4) building a strong balance sheet. Which of the following statement forms BEST describes the above statements?

a. visions
b. beliefs
c. missions
d. objectives
e. plans

Answer: (d) Difficulty: (3)

18. After visions and missions have been formulated, the company needs to translate these statements into operational ____________

a. options.
b. alternatives.
c. objectives.
d. plans.
e. tactics.

Answer: (c) Difficulty: (3)

19. A(n) ______________ is a course of action that explains how the enterprise will move from the business it is now in to the business it wants to be in, given its opportunities and threats and its internal strengths and weaknesses.

a. option
b. strategy
c. mission
d. plan
e. tactic

Answer: (b) Difficulty: (2)

20. ______________ means translating the strategy into actions and results.

a. Strategy formulation
b. Strategy development
c. Strategic alternative
d. Strategic implementation
e. Strategic control

Answer: (d) Difficulty: (3)

21. _______________ is a process to assess the firm's progress toward its strategic objectives and taking corrective action as needed.

a. Strategy quantification
b. Strategy development
c. Strategic modeling
d. Strategic implementation
e. Strategic control

Answer: (e) Difficulty: (2)

22. _______________ should ensure that all parts and members of the company are contributing in a useful way toward the strategy's implementation.

a. Strategy quantification
b. Strategy development
c. Strategic modeling
d. Strategic implementation
e. Strategic control

Answer: (e) Difficulty: (2)

23. ___________________ is the process of identifying the business of the firm today and the business it wants for the future, and then identifying the course of action or strategy it will pursue, given its opportunities, threats, and weaknesses.

a. Strategy quantification
b. Strategic planning
c. Strategic modeling
d. Strategic implementation
e. Strategic control

Answer: (b) Difficulty: (2)

24. All of the following strategic management tasks move the organization from where it is today to where it wants to be EXCEPT ___________________

a. defining the business.
b. developing a mission statement.
c. implementing the strategy.
d. setting strategic objectives.
e. formulation of a strategy to achieve strategic objectives.

Answer: (c) Difficulty: (2)

25. A strategic plan's purpose is to (pick the best general answer) ____________________

a. provide guidance for day-to-day decisions.
b. provide a barometer of each employee's performance.
c. balance the internal strengths and weaknesses of a company with its external opportunities and threats.
d. make a profit in all cases.
e. provide guidance for expansion.

Answer: (c) Difficulty: (3)

26. The purpose of situation analysis is to (pick the best general answer) ____________________

a. provide guidance for day-to-day decisions.
b. provide a barometer of each employee's performance.
c. balance the internal strengths and weaknesses of a company with its external opportunities and threats.
d. assess the internal features of and external forces on a company that will most directly affect its strategic options and opportunities.
e. provide guidance for expansion.

Answer: (d) Difficulty: (3)

27. A ____________________ purpose is to balance the internal strengths and weaknesses of a company with its external opportunities and threats.

a. situation analysis's
b. strategic plan's
c. operational plan's
d. functional plan's
e. tactical plan's

Answer: (b) Difficulty: (3)

28. The purpose of a(n) ____________________ is to assess the internal features of and external forces on a company that will most directly affect its strategic options and opportunities.

a. situation analysis
b. strategic plan
c. operational plan
d. functional plan
e. tactical plan

Answer: (a) Difficulty: (3)

29. All of the following are situational-analysis aids (developed by management experts) EXCEPT ________________

a. SWOT analysis.
b. conjoint analysis.
c. environmental scanning.
d. benchmarking.
e. examination of external forces in the company's environment.

Answer: (b) Difficulty: (2)

30. ________________ is a technique for summarizing a company's strengths, weaknesses, opportunities, and threats.

a. SWOT analysis
b. Environmental cataloging
c. Threat awareness charts
d. Defensive planning
e. Stratego simulation planning

Answer: (a) Difficulty: (1)

31. All of the following are illustrations of potential strengths EXCEPT ________________

a. market leadership.
b. strong research and development.
c. excess capacity for market.
d. high-quality products.
e. patents.

Answer: (c) Difficulty: (2)

32. All of the following are illustrations of potential threats EXCEPT ________________

a. diversification.
b. market saturation.
c. threat of takeover.
d. low-cost foreign competition.
e. slower market growth.

Answer: (a) Difficulty: (2)

33. The ________________ of an organization is defined as the set of forces with which that organization interacts. (Pick the best general answer.)

a. internal environment
b. competitive environment
c. hostile environment
d. international environment
e. external environment

Answer: (e) Difficulty: (2)

34. ___________________ is a systematic method of obtaining and compiling information about those environmental forces that might be relevant to the company's strategic planners.

a. Production scanning
b. External modeling
c. Systemic modeling
d. Environmental scanning
e. Environmental factoring

Answer: (d) Difficulty: (2)

35. In general, several key areas of the company's environment are "scanned" to identify the environmental forces or factors that may represent areas of concern, or represent opportunities or threats. All of the following represent these areas EXCEPT ____________________

a. technology.
b. economics.
c. statistics and abstracts.
d. social trends and demographics.
e. geography.

Answer: (c) Difficulty: (1)

36. Sometimes it's important to build plans and strategies based on knowledge of the very best that your competitors and others have to offer. The is commonly referred to as __________________

a. competitive modeling.
b. intelligence scanning.
c. brandmarking.
d. competitive scanning.
e. benchmarking.

Answer: (e) Difficulty: (2)

37. In the past few years, numerous manufacturers have gone to the Saturn car company plant in Springhill, Tennessee, in order to study (among other things) the human resource management practices that have created a unique partnership between management and labor. The above would best be described as an example of which of the following techniques?

a. competitive modeling
b. intelligence scanning
c. brandmarking
d. competitive scanning
e. benchmarking

Answer: (e) Difficulty: (2)

38. The basic benchmarking process adheres to all the following guidelines EXCEPT ____________

a. Studying the best practices is a two-way street; be willing to share information with others.
b. Studying the best practices of others is a good way to put intelligence-gathering procedures in place with their workers or hire them for your company.
c. Use employees who will actually be responsible for implementing any changes based on the procedures observed.
d. Focus on a specific problem and define it carefully.
e. Keep information that you receive confidential.

Answer: (b) Difficulty: (2)

39. The strategy(ies) that identifies the portfolio of businesses that will comprise the corporation, and the ways in which these businesses will relate to each other is called the ____________

a. corporate-level strategy.
b. competitive strategy.
c. differential strategy.
d. min-max strategy.
e. functional strategies.

Answer: (a) Difficulty: (1)

40. The strategy(ies) that identifies how to build and strengthen the business' long-term competitive position in the marketplace is called the ____________

a. corporate-level strategy.
b. competitive strategy.
c. differential strategy.
d. min-max strategy.
e. functional strategies.

Answer: (b) Difficulty: (1)

41. The strategy(ies) that identifies how Pizza Hut will compete against Domino's or Microsoft against Netscape is generally called the ____________

a. corporate-level strategy.
b. competitive strategy.
c. differential strategy.
d. min-max strategy.
e. functional strategies.

Answer: (b) Difficulty: (1)

42. The strategy(ies) that identifies the basic courses of action that each of the business' various departments will pursue to contribute to attaining the business' goals is called the ______________

a. corporate-level strategy.
b. competitive strategy.
c. differential strategy.
d. min-max strategy.
e. functional strategies.

Answer: (e) Difficulty: (2)

43. All of the following are considered to be corporate-level strategies EXCEPT ______________

a. concentration.
b. diversification.
c. vertical integration.
d. cost leadership.
e. investment reduction.

Answer: (d) Difficulty: (3)

44. A(n) ______________ (or single business) strategy means that the company will focus on one product or product line, usually in one market.

a. concentration
b. diversification
c. vertical integration
d. cost leadership
e. investment reduction

Answer: (c) Difficulty: (1)

45. ______________ is a growth strategy whereby the company acquires ownership of control of competitors who are competing in the same or similar markets with the same or similar products.

a. Vertical integration
b. Concentric growth
c. Horizontal integration
d. Diversification growth
e. Conglomerate diversification

Answer: (c) Difficulty: (3)

46. The strategy that is associated with expanding into other businesses through owning or controlling the inputs of the firm's processes and/or channels through which the firm's products or services are distributed is called ______________________

a. vertical integration.
b. concentric growth.
c. horizontal integration.
d. diversification growth.
e. conglomerate diversification.

Answer: (a) Difficulty: (3)

47. Diversification means a strategy of expanding into related or unrelated products or market segments. ____________ diversification means diversifying into other industries in such a way that a firm's lines of business still possess some kind of fit.

a. Related
b. Conglomerate
c. Retrenched
d. Universal
e. Parallel

Answer: (a) Difficulty: (2)

48. When Mobil Oil bought the Montgomery Ward retail chain, this diversification would be called ____________

a. related diversification.
b. retrenched diversification.
c. universal diversification.
d. parallel diversification.
e. conglomerate diversification.

Answer: (e) Difficulty: (2)

49. Investment reduction and defensive strategies are corrective actions taken to reduce the company's investments. When IBM dramatically reduced (downsized) its number of employees and closed a multitude of facilities, this was an example of which of the following reduction strategies?

a. Divestment
b. Bankruptcy
c. Liquidation
d. Capitulation
e. Retrenchment

Answer: (e) Difficulty: (2)

50. For many firms encountering rapid change, the ultimate strategic alliance is the ________________, "a temporary network of independent companies—suppliers, customers, and even rivals -- linked by information technology to share skills, costs, and access to one another's markets."

a. Internet
b. America Online
c. Lycos
d. virtual corporation
e. cabal

Answer: (d) Difficulty: (2)

51. When start-up company TelePad came up with an idea for a hand-held, pen-based computer, a(n) ____________ was its answer for breathing life into the idea. Several firms became involved: an industrial design firm in Palo Alto, California, designed the product; Intel helped with engineering details; several firms designed software; and a battery maker collaborated with TelePad to produce the power supply.

a. Internet connection
b. cartel
c. consortium
d. virtual corporation
e. cabal

Answer: (d) Difficulty: (2)

52. In formulating a corporate strategy, how do you decide which businesses to keep in the portfolio? The ______________ helps to identify the relative attractiveness of each of a firm's businesses. The method compares business growth rate and relative competitive position (market share) for each of the company's businesses.

a. BCG matrix
b. GE Business Screen
c. competitors curve
d. McDonald profit model
e. Compex model

Answer: (a) Difficulty: (3)

53. When measured on a matrix that uses business growth rate versus relative competitive position, ____________ are businesses in high-growth industries in which the company has a high relative market share. They usually require large infusions of cash to maintain growth.

a. Question marks
b. Dogs
c. Stars
d. Cash Cows
e. Comets

Answer: (c) Difficulty: (1)

54. When measured on a matrix that uses business growth rate versus relative competitive position, ____________ are businesses in high-growth industries, but with low market shares. Many times, cash diversion from richer divisions is required if these businesses are remain competitive.

a. question marks
b. dogs
c. stars
d. cash cows
e. comets

Answer: (a) Difficulty: (2)

55. ____________________ are often described as the generators of cash for the company's questionable or speculative businesses. Using the measurement of business growth rate versus market share, these businesses are low-growth industries that enjoy high relative market shares.

a. Question marks
b. Dogs
c. Stars
d. Cash cows
e. Comets

Answer: (d) Difficulty: (1)

56. The ____________ is a nine-cell matrix that can be used to analyze a business portfolio. Each company or unit is plotted into the appropriate cell according to its industry attractiveness and business unit position.

a. BCG matrix
b. GE Business Screen
c. competitors' curve
d. McDonald profit model
e. Compex model

Answer: (b) Difficulty: (3)

57. Whether a company decided to concentrate on a single business or to diversify into several different ones, it should develop competitive strategies for each of its businesses. Michael Porter has identified three primary strategies for doing this. Pick the best grouping of those strategies that are listed below.

a. Price leadership, segmentation, and focus
b. Cost leadership, differentiation, and focus
c. Cost leadership, segmentation, and focus
d. Price leadership, innovation, and leadership in technology
e. Cost leadership, innovation, and leadership in international affairs

Answer: (b) Difficulty: (3)

58. With respect to Michael Porter's primary strategies for meeting competition, ________________ is a prime example of a firm that is an industry cost leader. For example, its distribution costs are minimized through a satellite-based warehousing system.

a. K-Mart
b. Wal-Mart
c. 7-11
d. Jiffy Lube
e. IBM

Answer: (b) Difficulty: (2)

59. In a ________________ strategy, a firm seeks to be unique in its industry along some dimensions that are valued by buyers.

a. price leadership
b. cost leadership
c. segmentation
d. focus
e. differentiation

Answer: (e) Difficulty: (2)

60. A Pea in a Pod (a chain of maternity stores) concentrates on selling stylish clothes to pregnant working women. By concentrating on "working women's maternity clothes," the company is able provide a better offer than traditional department stores to their target customers. This would be an example of which of the following competitive strategies? (Pick the best and most general application.)

a. Price leadership
b. Cost leadership
c. Segmentation
d. Focus
e. Differentiation

Answer: (d) Difficulty: (3)

61. All of the following are forces driving industry competition (according to Michael Porter) <u>EXCEPT</u> ________________

a. potential entrants.
b. ethical relations.
c. suppliers.
d. buyers.
e. substitutes.

Answer: (b) Difficulty: (1)

62. At some point, each business' choice of competitive strategy is translated into supporting strategies for each of its departments to pursue. Which of the following best describes these strategies?

a. Functional strategies
b. Tactical strategies
c. Differentiation strategies
d. Corporate-level strategies
e. Profit strategies

Answer: (a) Difficulty: (2)

63. Formulating a strategic plan is just the first part of the strategic management process. The plan must be implemented. All of the following are implementation requirements EXCEPT ______________

a. Achieve a strategic fit between the strategy and the company's functional activities.
b. Delete non-profit components quickly and efficiently.
c. Leverage the firm's core competencies.
d. Provide the leadership from top management to implement the plan.
e. Provide a fit between corporate-level plans and operational plans.

Answer: (b) Difficulty: (3)

64. Consider McDonald's strategy for expanding overseas. Employees' knowledge and skills in new markets had to be expanded. New employees had to be recruited and trained. New reward systems had to be implemented that would appreciate the foreign cultures in new markets. Which of the following best typifies the McDonald's attempt to construct strategy?

a. Achieve a "strategic fit" between the strategy and the company's functional activities.
b. Downsize for excellence.
c. Create counter-cultures.
d. Appreciate top management mandates for success.
e. Move cautiously toward expansion.

Answer: (a) Difficulty: (1)

65. Successful companies (such as Wal-Mart) seem to be able to ____________ resources by concentrating them more effectively on key strategic goals.

a. fit
b. compound
c. defend
d. leverage
e. intensify

Answer: (d) Difficulty: (2)

66. ________________ (is) are defined as being "the collective learning in the organization, especially (knowing) how to coordinate diverse production skills and integrate multiple streams of technologies."

a. Leverage
b. Specialization
c. Core competencies
d. Leadership functions
e. Synergistic systems

Answer: (c) Difficulty: (2)

<u>True/False Short Answer</u>

67. According to management expert Peter Drucker, the primary task of top management is to ask the question "What is our business and what should it be?"

Answer: (True) Difficulty: (1)

68. ______________ is the process of identifying and pursuing the organization's mission by aligning the organization's internal capabilities with the external demands of its environment.

Answer: (Strategic management) Difficulty: (3)

69. The first step in the strategic management process is to set strategic objectives.

Answer: (False) Difficulty: (2)

70. Since Ferrari and Toyota both make cars, they would answer the question "What is our business?" in the same way.

Answer: (False) Difficulty: (1)

71. If Apex Elevator states that it wants "To provide a high reliability, error-free method for moving people and products up, down, and sideways within a building," they have formulated and stated a ______________ statement.

Answer: (mission) Difficulty: (3)

72. A vision is a "general statement of its (a company's) intended direction that evokes emotional feelings in organization members."

Answer: (True) Difficulty: (1)

73. The primary difference between a mission statement and a vision is in the degree of emotion solicited and direction that is provided.

Answer: (True) Difficulty: (2)

74. If a company tries to identify or communicate "who we are, what we do, and where we're headed," they are probably dealing with their ________________.

Answer: (mission statement) Difficulty: (3)

75. Strategic control is a course of action that explains how the enterprise will move from the business it is in now to the business it wants to be in, given its opportunities and threats and its internal strengths and weaknesses.

Answer: (False) Difficulty: (2)

76. There are several components of strategic planning. One of these components (which is often thought to play a central role), ______________, has the purpose of assessing the internal features of and external forces on a company that will most directly affect its strategic options and opportunities.

Answer: (situation analysis) Difficulty: (3)

77. SWOT stands for <u>S</u>trategies, <u>W</u>eaknesses, <u>O</u>pportunities, and <u>T</u>hrusts.

Answer: (False) Difficulty: (2)

78. L.L. Bean has a reputation for being a best-practice company because it expeditiously and courteously answers prospective customers' questions and fulfills their orders. If your company were to study their company (perhaps even visit their operational headquarters or distribution site) to gain ideas on how to improve your operations, this process of study and change would be called ________________.

Answer: (benchmarking) Difficulty: (3)

79. A concentration strategy means that the company focuses on one product or product line, usually in one market.

Answer: (True) Difficulty: (2)

80. Horizontal integration means owning or controlling the inputs of the firm's processes and/or the channels through which the firm's products or services are distributed.

Answer: (False) Difficulty: (3)

81. ____________ diversification means diversifying into products or markets that are not related to the firm's present businesses or to one another.

Answer: (Conglomerate) Difficulty: (2)

82. Within the BCG Matrix, the "question marks" are in high-growth industries, but with low market shares.

Answer: (True) Difficulty: (1)

83. When considering competitive strategies, if a firm seeks to be unique in its industry along some dimensions that are valued by buyers, the firm is said to be pursuing a focus strategy.

Answer: (False) Difficulty: (3)

84. A functional strategy is normally followed by only the top members of the organization.

Answer: (False) Difficulty: (2)

85. Examples of core competencies at Canon Corporation would be precision mechanics, fine optics, and micro-electronics.

Answer: (True) Difficulty: (1)

Essay

86. Define strategic management and then list and briefly identify the five strategic management steps as outlined by your text.

Answer:

Strategic management is defined as being the process of identifying and pursuing the organization's mission by aligning the organization's internal capabilities with the external demands of its environment.

The strategic management process consists of five steps:

a. Define the business and develop a mission statement. In this step the company must answer the question "What is our business?" In addition, the company should consider the vision and mission statements of the leaders in the organization.

b. Translate the mission into strategic objectives. Here the company must get down to work and move the mission from an idea to a reality. In other words, the mission becomes operational strategic objectives.

c. Formulate a strategy to achieve the strategic objectives. A strategy is a course of action that explains how the enterprise will move from the business it is now in to the business it wants to be in (as stated in the mission), given its opportunities and threats and its internal strengths and weaknesses.

d. Implement the strategy. All functions within the firm should be brought together to give life to the strategies devised by management.

e. Evaluate performance and correct as required. Use strategic control to monitor progress toward objectives and make appropriate corrections.

Difficulty: (2)

87. Identify what SWOTs stands for. Next, provide an example that demonstrates each of the components of SWOTs. Be specific in your comments.

Answer:

SWOTs stands for Strengths, Weaknesses, Opportunities, and Threats.

Students may provide a variety of answers for the second part of this question. However, the instructor should judge the validity of the response against the items listed in Figure 6.4 (page 183). For example Strengths = market leadership; Weaknesses = large inventories; Opportunities = new overseas markets; and Threats = market saturation.

Difficulty: (1)

88. The text lists five primary corporate-level strategies that the strategic planner might consider using to solve problems of the firm. These strategies help the firm to identify what businesses to be in and how these businesses should relate to each other. List and then discuss each of these strategies. Lastly, since most students seem to prefer one of these strategies above the others, pick one and tell what you find particularly attractive about your choice.

Answer:

a. Concentration strategy. A concentration strategy (or single business strategy) means the company focuses on one product or product line, usually in one market. Student answers should explain the concept and demonstrate the sub-strategy components of market penetration, geographic expansion, product development, and horizontal integration.

b. Vertical integration. Instead of staying in one business, a firm can expand into other businesses through a strategy of vertical integration. In this strategy the firm owns or controls the inputs to the firm's processes and/or the channels through which the firm's products or services are distributed.

c. Diversification. Diversification means a strategy of expanding into related or unrelated products or market segments. Diversification can take on the forms of related diversification (where there is some form of fit) or conglomerate diversification (where expansion is in unrelated areas).

d. Status-quo strategies. Stability or status-quo strategy indicates that the organization is satisfied with its rate of growth and product scope.

e. Investment reduction strategies. Investment reduction or defensive strategies are corrective actions taken to reduce the company's investments in one or more of its lines of business. The strategies can be retrenchment (downsizing) or divestment (selling or liquidating businesses).

Difficulty: (3)

89. In formulating a corporate strategy, an often-asked question is "How do you decide which businesses to keep in the portfolio?" One of the techniques that can be applied to the quest for an answer is the Boston Consulting Group's Matrix. Describe the method and each of its component parts. After describing the parts (or cells) of the BCG Matrix, think of an illustration for each of the cells (indicate how the examples that you chose fit the cells).

Answer:

The BCG Matrix helps to identify the relative attractiveness of each of a firm's businesses. The method compares business growth rate and relative competitive position (market share) for each of the company's businesses.

a. Stars: Stars are businesses in high-growth industries in which the company has a high relative market share. Intel's microprocessor business has a high relative growth rate. Star businesses usually require large infusions of cash to sustain growth.

b. Question marks: Question marks are businesses in high-growth industries, but with low market shares. An example would be the computer business started by Exxon Oil several years ago. In order to survive in these businesses, the company must divert large sums of cash from its other businesses to fend off competitors or get out of the business altogether.

c. Cash cows: Cash cows are businesses in low-growth industries that enjoy high relative market shares. Their being in low-growth, unattractive industries argues against making large cash infusions in these businesses. However, their high market share generally allows them to generate high sales and profits for years, even without much new investment. Microsoft's MS-DOS is an example of the cash cow in the rapidly expanding computer software business.

d. Dogs: Finally, dogs are low-market share businesses in low-growth, unattractive industries. Having a low-market share puts the business in jeopardy relative to its larger competitors. Dogs can quickly become "cash traps." A marginal venture in the classical music recording field would be an example.

Note: If the instructor wishes to shorten this question, delete the example assignment.

Difficulty: (2)

90. Strategy expert Michael Porter argues that the way a company competes—its competitive strategy—depends largely on how intense the competition is in an industry. He believes that competition intensity reflects five competitive forces. List and then analyze each of these.

Answer:

a. Threat of entry (new entrants): Intensity of industry competition depends first on the threat of new entrants to the industry. In general, the more easily new competitors can enter the business, the more intense the competition.

b. Intensity of rivalry among existing competitors: Rivalry manifests itself in tactics like price competition, advertising battles, and increased customer service. In rivalry, some industries appear to be more intense and warlike. Cutthroat practices are common.

c. Pressure from substitute products: Intensity also depends on substitute products. The more substitutes available, the more competitive the industry. When substitutes are reduced (as in certain patented medicine areas), the industry is more attractive and less cutthroat to current competitors.

d. Bargaining power of buyers: The buyer is a factor. If buyer's purchase large volumes, they are more powerful. Similarly, if the products are standard or undifferentiated or if the buyers face few switching costs, they tend to have more control over the suppliers.

e. Bargaining power of suppliers: Suppliers can influence the process by threatening to raise prices or reduce quality. Suppliers are more powerful if they are dominated by just a few suppliers and are more concentrated. Few substitutions contribute to supplier power.

Difficulty: (3)

Chapter 7
FUNDAMENTALS OF ORGANIZING

INTRODUCTION

This chapter includes the definition of organizing, a description of the basic alternatives for organizing departments, identification of the types of authority in organizations, demonstration of how authority can be delegated, and discussion of decentralization in organizations.

LECTURE OUTLINE

What's Ahead?

What Is Organizing?

Creating Departments
- Creating Departments Around Functions
- Creating Departments Around Products
- Creating Departments Around Customers
- Creating Departments Around Marketing Channels
- Creating Departments Around Geographic Areas
- Creating Matrix Organizations
- Departmentalization in Practice: A Hybrid

Achieving Coordination
- The Nature and Purpose of Coordination
- Techniques for Achieving Coordination

Authority in Organizations
- Sources of Authority
- Power and Politics At Work
- Line and Staff Authority
- Line and Staff Organizations
- The Delegation Process
- The Decentralized Organization

Tall and Flat Organizations and the Span of Control
- Flat Versus Tall Organizations
- The Span of Control

Summary

Case: The New Visa Organization
You Be the Consultant: Organizing KnitMedia

ANNOTATED OUTLINE

What's Ahead?

When Starbuck's Coffee operated a few stores on the West Coast the organizational structure was fairly simple. Today, with over 1000 Starbucks (and growing) in the United States, overseas expansion, and departments selling to airlines, supermarkets, and other institutional customers, the organizational structure has become complex and is an important factor in the firm's strategy for the future.

What Is Organizing?

Organizing is the arranging of enterprise activities in such a way that they systematically contribute to the goals of the enterprise. An organization consists of people whose specialized tasks are coordinated to contribute to the goals of the organization.

Notes:

Creating Departments

Departmentalization is the process through which the activities of an enterprise are grouped together and assigned to managers. The organization-wide division of work is departmentalization.

Notes:

Creating Departments Around Functions. Functional departmentalization involves grouping activities around basic functions like manufacturing, sales, and finance. This form of organization is simple and logical, it avoids duplication of effort, it enables managers to specialize and use expertise, training of managers is easier, and managers generally work with information relating to their function which makes it easier for executives to exercise tight control if they so desire. Disadvantages of the functional form include placing large responsibility on the president, the development of too many specialized managers at the expense of developing generalists, and coordination activities may be too difficult for one person.

Notes:

Creating Departments Around Products. Departments are organized for each of the company's products or for each family of products. Arranging departments around products is called divisionalization and each department can manage all the activities needed to develop, manufacture, and sell a particular product or product line. Divisionalization includes a single manager overseeing all the functions required to produce and market each product. This arrangement is appropriate when quick decisions and flexibility are paramount. Disadvantages of divisionalization include expensive duplication of effort, diminished control by top management, and the requirement for more general managers that are harder to train and develop.

Notes:

Creating Departments Around Customers. Customer departmentalization is similar to divisionalization because departments are organized to serve the needs of specific customers. The main divisions are organized to serve the needs of particular customers. A manager can give his/her undivided attention to a customer or group of customers, resulting in faster, more satisfactory service to each customer. An important disadvantage is duplication of effort.

Notes:

Creating Departments Around Marketing Channels. Marketing channel departmentalization has top-level departments organized around each of the marketing channels of the firm. A marketing channel is the conduit through which a manufacturer distributes its products to ultimate customers. In marketing channel departmentalization, the same product is typically marketed through two or more channels. One department usually manufactures the product for all the other marketing channel departments.

Notes:

Creating Departments Around Geographic Areas. Geographic departmentalization includes separate departments organized for each of the territories in which the enterprise does business. An advantage is that one self-contained department is charged with focusing on the needs of its particular buyers, here in its geographic area.

Notes:

Creating Matrix Organizations. A matrix organization is defined as the superimposition of one or more forms of departmentalization on top of an existing form. One type includes product departments superimposed over a functional departmentalization. Problems with matrix organizations include power struggles and conflicts, lost time in intragroup meetings, and excessive overhead because of hiring more managers and secretaries.

Notes:

Departmentalization in Practice: A Hybrid. Most organizations use several forms of departmentalization and are called hybrids. For example, top management might use functional departments for production, sales, and finance. It might organize the sales department by geographic areas.

Notes:

Achieving Coordination

The Nature and Purpose of Coordination. Coordination is the process of achieving unity of action among interdependent activities. It is required when two or more individuals, groups, or departments must work together to achieve a common goal.

Notes:

The Nature and Purpose of Coordination. Whenever two or more independent activities seek to work together to achieve a common goal it is necessary that their actions be as one. When the Egyptian pyramids were being constructed there were huge blocks of stone moved by manpower alone. It was necessary for 5,000 workers to pull together, all at the same time and all in the same direction. Without coordination much effort can be wasted and the task is in jeopardy.

Notes:

Techniques for Achieving Coordination. Techniques for achieving coordination include mutual adjustment that relies on interpersonal communication to coordinate in both the simplest and most complex situations. Rules and procedures can be used for coordinating routine, recurring activities. Direct supervision through using the hierarchy can achieve coordination by having one person coordinate the work of others and monitoring results. Divisionalization can facilitate coordination as the president must coordinate the functions of the business. Staff assistants can be hired to make the job of coordinating subordinates easier. Liaisons can be used to coordinate between two departments when the volume of contacts increases significantly. Committees may be appointed to coordinate activities. Independent integrators can be used to coordinate the activities of several interdependent departments. Integrators differ from liaisons because they are independent of the departments they coordinate. Standardized targets, skills, or shared values can achieve coordination by standardizing the efforts of employees.

Notes:

Authority In Organizations

Sources of Authority. Authority is the right to take action, make decisions, and direct the work of others. Authority may be derived from a variety of sources, and Chester Barnard argued that for orders to be carried out, they must lie within the subordinate's "zone of acceptance." Getting employee acceptance is increasingly important today because of emphasis on empowered workers and team-based organizations.

Notes:

Power and Politics at Work. The five bases of power are identified, including reward power, coercive power, legitimate power, referent power, and expert power. The acquisition and use of power in organizations is called politics. Employees can win at office politics by controlling vital information, hooking their career to a star, minimizing power vacuums, and opening communication channels.

Notes:

Line and Staff Authority. Line managers are always in charge of essential activities such as sales, and they are authorized to issue orders to subordinates down the chain of command. Staff managers generally cannot issue orders down the chain of command except in their own departments. Staff managers may have functional authority meaning that they can issue orders down the chain of command within the very narrow limits of their authority.

Notes:

Line and Staff Organizations. Most organizations have both line and staff departments. Line-staff conflict refers to disagreements between a line manager and the staff manager who is giving advice.

Notes:

The Delegation Process. Delegation is the pushing down of authority from supervisor to subordinate. The assignment of responsibility to get a job done goes along with the delegation of authority to accomplish it. Empowerment means authorizing and enabling workers to do their jobs.

Notes:

The Decentralized Organization. A decentralized organization is one in which authority for most decisions is delegated to the department heads while control over essential company-wide matters is maintained at the headquarters office. The more decisions and authority are delegated from the president to subordinates, the more decentralized the organization is.

Notes:

Tall and Flat Organizations and the Span of Control

Flat Versus Tall Organizations. Jack Welch had to make significant changes at GE after he became CEO. The chain of command was draining resources because of duplication of work by managers. Welch flattened the organization and the chain of command. The company is much more responsive and its corporate results have reflected its new effectiveness. These types of restructuring efforts have occurred at many organizations in recent years. Restructuring, however, can lead to problems with span of control for managers.

Notes:

The Span of Control. The span of control is the number of subordinates reporting directly to a supervisor. There is a correlation between the number of people reporting to a manager and the number of management levels in an organization. While tall organizations with narrow spans of control may require close supervision, they can also backfire by slowing decision making. As the rate of technological change and new-product introductions has increased, many more firms have opted for flatter structures and wider spans.

Notes:

Summary

Notes:

Case: The New Visa Organization

1. Management now means more collaboration, authority, initiative, decision making, and profits pushed to the outside of the organization as a result of the decentralization effort. A new organization was designed to carry out the new strategy.

2. Command and control organizations are not very effective in a highly competitive product market requiring constant innovation and rapid response to changes in the market. For these and other reasons, this type of organization has been on the decline. It may still function well if a firm is not in a highly competitive environment and it dominates the market, but how long will the situation last?

3. This is possible as organizations decentralize and empower employees. The proliferation of ESOPS also may result in more of this in the future. Of course, an organization with too many chiefs and not enough warriors may have its problems, also.

You Be the Consultant: Organizing KnitMedia

1. KnitMedia appears to have a hybrid organizational structure with product and geographic departmentalization.

2. There could be some confusion regarding the roles of the three top-level managers. If this were the case, it would be difficult to assign reporting relationships to the top.

3. The three people at the top could be an executive committee. The organization could be flattened and duplication of effort eliminated. Perhaps a matrix organization based on marketing and production/service operations would work well. If not, the firm might want to continue in the same mode with more clarity in reporting to top management as a committee. If the organization becomes too unwieldy as the firm continues to grow, it will have to be flattened to respond well to customers globally by making decisions more quickly and efficiently. Whatever structure the firm uses, it must follow and complement the organizational strategies.

End of Chapter Questions for Chapter 7

Multiple Choice

1. Arranging the activities of the enterprise in such a way that they systematically contribute to the enterprise's goals is called ____________________

a. the chain of command.
b. delegation.
c. organizing.
d. departmentalization.
e. coordination.

Answer: (c) Difficulty: (2)

2. A(n) ________________ consists of people whose specialized tasks are coordinated to contribute to the enterprise's goals.

a. bureaucracy
b. committee
c. cell
d. corporation
e. organization

Answer: (e) Difficulty: (2)

3. The most common and usual way of depicting an organization is with a(n) ____________
a. bar chart.
b. pie chart.
c. schematic.
d. organization chart.
e. Gantt chart.

Answer: (d) Difficulty: (1)

4. The structure of the organization, specifically, the title of each manager's position and, by means of connecting lines, who is accountable to whom and who is in charge of what department, is shown by the ____________ chart.

a. organization
b. Gantt
c. structural
d. time-line
e. procedural

Answer: (a) Difficulty: (1)

5. The organization chart shows what is sometimes called the scalar chain or the line of authority between the top of the organization and the lowest positions in the chart. The common name for this relationship is the ____________. Formally, it represents the path a directive should take in traveling from the president to employees at the bottom of the organization chart or from employees at the bottom to the top of the organization.

a. Gantt chart.
b. military diagram chart.
c. communication chart.
d. chain of command.
e. bureaucracy path.

Answer: (d) Difficulty: (2)

6. All of the following terms describe (or are used to provide descriptions of) the path a directive should take in traveling from the president to employees at the bottom of the organization chart or from employees at the bottom to the top of the organization EXCEPT ______________

a. chain of command.
b. scalar chain.
c. Gantt chart.
d. line of authority.
e. formal organizational relationships.

Answer: (c) Difficulty: (3)

7. The organization chart shows all of the following EXCEPT ______________

a. the informal organization.
b. the layers of management.
c. the line and staff relationships.
d. the departmentalization (if any).
e. the management hierarchy.

Answer: (a) Difficulty: (2)

8. A description of the ______________ organization includes those casual, habitual contacts, communications, and ways of doing things that employees develop within almost all organizations.

a. formal
b. bureaucratic
c. informal
d. communication
e. international

Answer: (c) Difficulty: (2)

9. A salesperson might develop the habit of calling a plant production supervisor to check on the status of an order. The salesperson might find this quicker than adhering to the chain of command, which would entail having the sales manager check with the plant manager, who in turn checks with the supervisor. This scenario would be an example of a(n) ______________ organization relationship.

a. formal
b. trade
c. communication
d. bureaucratic
e. informal

Answer: (e) Difficulty: (3)

10. ________________ is the process through which an enterprise's activities are grouped together and assigned to managers; it is the organization-wide division of work.

a. Entrepreneurship
b. Departmentalization
c. Compartmentalization
d. Bureaucracy
e. Control

Answer: (b) Difficulty: (2)

11. Departmentalization is the process through which an enterprise's activities are grouped together and assigned to managers. All of the following terms are used todesignate these logical groupings of activities EXCEPT (pick the least likely designation) ________________

a. departments.
b. employees.
c. divisions.
d. units.
e. sections.

Answer: (b) Difficulty: (3)

12. Under normal circumstances, departments can be created around all of the following activities or areas EXCEPT (pick the least likely) ________________

a. functions.
b. products.
c. purchasing.
d. customers.
e. marketing channels.

Answer: (c) Difficulty: (2)

13. When departments are organized around basic functions such as manufacturing, sales, and finance, this is an illustration of ________________ departmentalization.

a. functional
b. product
c. customer
d. marketing channel
e. geographic

Answer: (a) Difficulty: (1)

14. If a bank organizes its departments around the basic activities of operations, control, loans, and retail customer activity, then the bank is organized by ____________ departmentalization.

a. functional
b. product
c. customer
d. marketing channel
e. geographic

Answer: (a) Difficulty: (2)

15. Beneath the Chairperson and the Managing Board within a company are four Director positions (Export Sales Director, Finance Director, Domestic Sales Director, and Production Director). Each of these Directors manages his/her own departments separately. Considering the relationships just described, the company is most likely to be departmentalized based on ______________ relationships.

a. product
b. functional
c. customer
d. marketing channel
e. geographic

Answer: (b) Difficulty: (2)

16. Functional departmentalization has a series of advantages and disadvantages. All of the following would be considered to be advantages of functional departmentalization EXCEPT ______________

a. the chief executive normally has little responsibility in this organizational form.
b. managers functionally specialize and, therefore, are more efficient.
c. there is less duplication of effort than in other types of organization.
d. there are increased returns to scale.
e. it simplifies training.

Answer: (a) Difficulty: (3)

17. Functional departmentalization has a series of advantages and disadvantages. All of the following would be considered to be disadvantages of functional departmentalization EXCEPT ______________

a. Responsibility for overall performance lies only with the chief executive.
b. Training is simplified.
c. It can overburden the chief executive.
d. It reduces the attention paid to specific products, customers, markets, or areas.
e. It results in functionally specialized managers rather than general managers.

Answer: (b) Difficulty: (3)

18. Gillette Company is organized around the ______________ departmentalization format. For example, they have divisions for drugs and pharmaceuticals, personal care, and stationery products. Internationally, they are divided among two divisions (the international division and diversified operations).

a. functional
b. product
c. customer
d. marketing channels
e. geographic

Answer: (b) Difficulty: (1)

19. Arranging departments around products is often called __________________. This means that the firm's major departments are organized so that each can manage all the activities needed to develop, manufacture, and sell a particular product or product line.

a. departmentalization.
b. internationalization.
c. globalization.
d. divisionalization.
e. unitization.

Answer: (d) Difficulty: (2)

20. To Bill Harris, executive vice president of the software company Intuit, it became clear that the bigger the company got, the more that being functionally organized was a liability. It was time for a change. The basic goal was to flatten the organization. Each business unit would be the size that Intuit had been before its dramatic growth spurt and would focus on one core product or market. This would be an example of which of the following forms?

a. Departmentalization
b. Internationalization
c. Globalization
d. Divisionalization
e. Unitization

Answer: (d) Difficulty: (2)

21. Product departmentalization has a series of advantages and disadvantages. All of the following would be considered to be advantages of product departmentalization EXCEPT __________

a. One unit is responsible for giving continuous, undivided attention to the product.
b. Part of the burden is lifted from the shoulders of the top manager.
c. Performance is more easily identified and judged.
d. Efficiency is high because customers are targeted by only one division.
e. It is responsive to the unique needs of the product.

Answer: (d) Difficulty: (3)

22. Product departmentalization has a series of advantages and disadvantages. All of the following would be considered to be advantages of product departmentalization EXCEPT ___________

a. It increases duplication of effort, and is perhaps reduced efficiency.
b. Finding and training people to head each division is difficult.
c. Divisions have trouble giving continuous, undivided attention to products because so many are produced.
d. The top manager may loose control.
e. The top manager may have difficulty coordinating divisions.

Answer: (c) Difficulty: (2)

23. An example of a company that uses ________________ departmentalization is Grayson Steel Company. Separate departments are organized around ________________. Products go to aerospace, metal and chemicals, and international divisions. (Pick the most likely structural format)

a. functional; functions
b. product; products
c. customer; customers
d. marketing channels; marketing channels
e. geographic; geographic regions

Answer: (c) Difficulty: (1)

24. The primary advantage of the customer departmentalization format is that ____________

a. the manager is charged with giving undivided attention to a customer or group of customers.
b. there is little duplication of effort.
c. there is high corporate efficiency.
d. finding and training people to lead the division is easy if the manager looks outside the company.
e. coordination is usually not a problem.

Answer: (a) Difficulty: (2)

25. The main disadvantage of the customer departmentalization format is that

a. the top manager is overburdened.
b. performance is difficult to judge.
c. it is hard to give undivided attention to clients in this form.
d. duplication of effort is increased.
e. the top manager is able to maintain very tight control.

Answer: (d) Difficulty: (2)

26. A(n) ______________ is the conduit (wholesaler, retailer, etc.) through which a manufacturer distributes its products to its ultimate consumers.

a. physical distribution system
b. channel of producers
c. channel of distribution
d. cartel
e. producer network

Answer: (c) Difficulty: (1)

27. The primary difference between marketing-channel departmentalization and customer departmentalization is that, under the marketing-channel department arrangement, the department is ____________________

a. responsible for manufacturing and selling its own product to its own customers.
b. manufacturing the product while other departments distribute the product to their individual customers or channels.
c. not able to cater to unique customer needs.
d. generating profits that are generally lower with this form.
e. usually not as efficient as with the customer departmentalization format.

Answer: (b) Difficulty: (2)

28. If a company has a Managing Director for Australia, a Regional Manager for Europe, a Regional Manager for Asia, and a Controller, the company is best described as being organized in a ________________ format.

a. functional
b. product
c. customer
d. marketing channels
e. geographic

Answer: (e) Difficulty: (1)

29. As is the case with other types of divisional organizations, the main advantage of territorial (geographic) departmentalization is that ________________

a. there is no duplication of effort.
b. the manager is charged with giving undivided attention to a customer or group of customers—in this case, those in a particular geographic area.
c. there is high corporate efficiency.
d. finding and training people to lead the division is easy if the manager looks outside the company.
e. coordination is usually not a problem.

Answer: (b) Difficulty: (2)

30. A _______________ organization is defined as the superimposition of one or more forms of departmentalization on top of an existing form.

a. parallel
b. convex
c. concave
d. matrix
e. multifaceted

Answer: (d) Difficulty: (2)

31. Within a _______________ organization, a project structure is often superimposed over a functional organization.

a. parallel
b. convex
c. concave
d. matrix
e. multifaceted

Answer: (d) Difficulty: (2)

32. Consider the following example: A company's automotive products division is functionally organized, with departments for functions like production, engineering, and personnel. But superimposed over this functional departmentalization are three product groups—-for the Ford project, the GM project, and the Chrysler project. Each of these groups has its own manager. One or more employees from each functional department (like production and engineering) is temporarily assigned to each project. This would best be described as being a _______________ organization format.

a. parallel
b. convex
c. matrix
d. dual
e. multifaceted

Answer: (c) Difficulty: (2)

33. Matrix departmentalization can help some companies gain advantages over competition, however, the matrix format also has some serious disadvantages. All of the following would be classed as disadvantages EXCEPT _______________

a. the inability of a department to devote its undivided attention to the needs of its own project, product, or customer.
b. power struggles.
c. conflicts that naturally occur.
d. lost time.
e. excessive overhead.

Answer: (a) Difficulty: (2)

34. Matrix departmentalization can help some companies gain advantages over competition. It is thought that a large organization can gain the advantages that are characteristic of smaller firms by using this format. Consider the phrases listed below. Pick out the phrase that would be the chief advantage of the matrix departmentalization format.

a. Reduction in power struggles
b. Reduction in conflicts
c. Time efficiency
d. The ability of a department to devote its undivided attention to the needs of its own project, product, or customer (even though the project may be temporary)
e. Reduced overhead

Answer: (d) Difficulty: (3)

35. The process of achieving unity of action among interdependent activities is called __________

a. processing.
b. hybridization.
c. control.
d. synthesis.
e. coordination.

Answer: (e) Difficulty: (1)

36. Whenever two or more interdependent individuals, groups, or departments must work together to achieve a common goal, ________________ is necessary.

a. participation
b. hybridization
c. control
d. synthesis
e. coordination

Answer: (e) Difficulty: (2)

37. Through the work of Jay Galbraith and Henry Mintzberg, a set of techniques have been devised to aid the process of coordination at the managerial level. All of the following are ideas that come from the Galbraith and Mintzberg lists EXCEPT ____________________

a. coordination through mutual adjustment.
b. coordination through intimidation and coercion.
c. coordination through rules or procedures.
d. coordination through divisionalization.
e. coordination through committees.

Answer: (b) Difficulty: (3)

38. Through the work of Jay Galbraith and Henry Mintzberg, a set of techniques have been devised to aid the process of coordination at the managerial level. One of the coordination techniques is called ________________. It relies on interpersonal communication (whether the task is simple or complex). For example, if two people are moving a heavy log, coordination could be achieved by having one person count "one, two, three, lift," at which time both people lift the log in unison.

a. coordination by rule
b. coordination by supervision
c. coordination through liaisons
d. coordination by mutual adjustment
e. coordination by shared values

Answer: (d) Difficulty: (3)

39. The coordination of effort needed for complex tasks such as a barn-raising is the result of a communication process called ________________.

a. coordination by rules.
b. coordination by supervision.
c. coordination through liaisons.
d. coordination by mutual adjustment.
e. coordination by shared values.

Answer: (d) Difficulty: (3)

40. Through the work of Jay Galbraith and Henry Mintzberg, a set of techniques have been devised to aid the process of coordination at the managerial level. One of the coordination techniques is called coordination through independent integrators. An independent integrator is an individual or group that (pick the best and most general description) ________________

a. asks questions.
b. is the same as a liaison.
c. coordinates the activities of several interdependent departments.
d. facilitates the purchasing process.
e. audits the organization.

Answer: (c) Difficulty: (2)

41. Firms can achieve coordination by standardizing the efforts of their employees. This can be accomplished in all the following ways EXCEPT (pick the least likely alternative) __________

a. standardizing the goals the employees are to reach.
b. standardizing the skills so employees can work as a team.
c. standardizing the targets the employees are to reach.
d. standardizing the pay to produce more incentive for performance.
e. facilitate coordination by creating shared values among employees.

Answer: (d) Difficulty: (3)

42. ____________ is the right to take action, to make decisions, and to direct the work of others.

a. Control
b. Coordination
c. Authority
d. Leadership
e. Sponsorship

Answer: (c) Difficulty: (2)

43. Authority in an organization is derived from several sources. All of the following would be considered to be those sources <u>EXCEPT</u> (pick the least likely) ________________

a. a person's position.
b. a person's craftiness.
c. a person's rank.
d. a person's personal traits.
e. a person's expertise in the area.

Answer: (b) Difficulty: (2)

44. Some management writers argue that authority must come from the bottom up and be based on the subordinate's willingness to be led. Theorist Chester Barnard was an early proponent of this view. Barnard argued that, for orders to be carried out, they must lie within the subordinate's ________________ or problems would occur and there would be a breakdown in authority.

a. zone of compliance
b. zone of response
c. zone of acceptability
d. zone of rejection
e. zone of indifference

Answer: (c) Difficulty: (2)

45. Though many of the terms that are associated with the concept of authority seem to have the same meaning. There are some important differences. For example, _______________ can be defined as the potential for producing an effect—the effect of somehow getting someone or something to take some action.

a. influence
b. power
c. authority
d. leadership
e. coercion

Answer: (a) Difficulty: (3)

46. Though many of the terms that are associated with the concept of authority seem to have the same meaning there are some important differences. For example, ______________ can be defined (by Max Weber) as "the probability that one actor within a social relationship will be in a position to carry out his own will despite resistance, regardless of the basis on which this probability rests."

a. influence
b. power
c. authority
d. leadership
e. coercion

Answer: (b) Difficulty: (3)

47. Though many of the terms that are associated with the concept of authority seem to have the same meaning there are some important differences. For example, ______________ can be defined as referring to a person's potential for influencing others, but the word has the implicit notion of legitimacy or ethical sanctification.

a. influence
b. power
c. authority
d. leadership
e. coercion

Answer: (c) Difficulty: (3)

48. Power, in organizational behavior, does not have to be "legitimate," and in fact, social scientists distinguish among all of the following as bases or sources of power EXCEPT: (pick the least likely) ____________

a. reward power.
b. synthetic power.
c. coercive power.
d. referent power.
e. legitimate power.

Answer: (b) Difficulty: (2)

49. Social scientists have distinguished at least five bases or sources of power in an organization. A person has ______________ if he or she can significantly influence whether a person gets more money (as in a raise) or can reduce a person's poor working conditions (give the person a better office or working climate).

a. reward power
b. expert power
c. coercive power
d. referent power
e. legitimate power

Answer: (a) Difficulty: (1)

50. Social scientists have distinguished at least five bases or sources of power in an organization. A person has ______________ over another if there is a real or imagined expectation on the part of the latter that he or she will be punished for failing to conform to the powerful person's attempts at influence.

a. reward power
b. expert power
c. coercive power
d. referent power
e. legitimate power

Answer: (c) Difficulty: (2)

51. When bosses use fear, intimidation, and threats as a means to get conformity among their employees, they are using ______________

a. reward power.
b. expert power.
c. coercive power.
d. referent power.
e. legitimate power.

Answer: (c) Difficulty: (1)

52. If you agree to join a company's sales department, you are in essence accepting the the right of the sales manager to assign you work and review your progress. In an organizational sense, you are respecting and conforming to the direction of the position (not necessarily the person that holds that position). This form of power is called ______________

a. reward power.
b. expert power.
c. coercive power.
d. referent power.
e. legitimate power.

Answer: (e) Difficulty: (2)

53. ______________ is roughly synonymous with what has been called authority and refers to the power that stems from a person feeling that someone has a right to influence him or her and that he or she has an obligation to accept this influence.

a. Reward power
b. Expert power
c. Coercive power
d. Referent power
e. Legitimate power

Answer: (e) Difficulty: (1)

54. ________________ stems from a person's identification with another.

a. Reward power
b. Expert power
c. Coercive power
d. Referent power
e. Legitimate power

Answer: (d) Difficulty: (2)

55. Using the following example: The new presidential assistant says of the president, "I want to be like that person, therefore, I shall behave or believe as the president does." This would be an example of which of the following forms of power?

a. Reward power
b. Expert power
c. Coercive power
d. Referent power
e. Legitimate power

Answer: (d) Difficulty: (2)

56. ________________ derives from a person being viewed as the top person in some area and on whom others must therefore depend for advice and counsel.

a. Reward power
b. Expert power
c. Coercive power
d. Referent power
e. Legitimate power

Answer: (b) Difficulty: (1)

57. The way the organization is structured has a big influence on who will have ________________. For example, ________________ in a company often stems from a person's position in the communications network and from that person's ability to control access to coveted information.

a. reward power
b. expert power
c. coercive power
d. referent power
e. legitimate power

Answer: (b) Difficulty: (2)

58. In organizations, the acquisition and use of power is called ___________.

a. gamesmanship.
b. jockeying.
c. up-the-organization.
d. organizational culture.
e. politics.

Answer: (e) Difficulty: (3)

59. When examining relationships shown on organizational charts, managers distinguish between different forms of authority. ______________ managers generally cannot issue orders down the chain of command (except in their own departments). They usually only assist and advise. (pick the most accurate term) ______________

a. Staff
b. Senior
c. Line
d. Specialist
e. Head

Answer: (a) Difficulty: (2)

60. Consider the following example: A production manager may want to use a particular personnel test. However, the HR manager insists that the test not be used. This would be generally be described as a ______________ conflict. (Pick the best organizational response listed.)

a. line-line
b. staff-staff
c. manager
d. line-staff
e. power

Answer: (d) Difficulty: (3)

61. When authority is pushed down from supervisor to subordinate, it is called ______________

a. power-brokering.
b. blocking.
c. delegation.
d. leadership.
e. coordination.

Answer: (c) Difficulty: (2)

62. Consider the following example: At Toyota the employees are not only given the authority to solve problems on the assembly line, they are also given the training, tools, and management support required to enable them to solve their problems. This process is called ________________

a. empowerment.
b. improvement.
c. shifting responsibility.
d. diffusion.
e. Intrapreneurship.

Answer: (a) Difficulty: (2)

63. Organizations can be organized into many forms. If you were to construct an organization in which authority for most decisions was delegated to the department heads (usually the heads of product divisions), while control over essential company-wide matters was maintained at the headquarters office, you would have designed a ______________

a. centralized organization.
b. random-box organization.
c. "clover leaf" organization.
d. decentralized organization.
e. multi-functional organization.

Answer: (d) Difficulty: (2)

64. The more decisions and the more areas in which authority is delegated from the president to his or her subordinates, the more ______________ the organization is.

a. centralized
b. random-box
c. "clover leaf"
d. decentralized
e. multi-functional

Answer: (d) Difficulty: (2)

65. Economic historian Alfred Chandler discovered in his classical study of organizations that ____________. He found that a company's organizational structure had to fit or be appropriate to the nature of its strategy. Seeing the relationship was extremely important. (Pick the most appropriate statement that illustrates Chandler's thoughts.)

a. strategy follows structure
b. structure follows strategy
c. product strategy rules all forms of structure
d. communication strategy usually runs contrary to formal structure
e. centralization is the best structure

Answer: (b) Difficulty: (3)

66. If a manager attempts to streamline the organization by stripping away unnecessary levels of "bureaucracy," the manager is said to be ____________ the organization. (Pick the most descriptive term.)

a. fattening
b. flattening
c. trimming
d. peaking
e. circumventing

Answer: (b) Difficulty: (1)

67. If a manager attempts to push authority down to the lower levels of the organization, restructure, shorten chains of command, and empower employees, then the organization is most likely going through a ____________ process.

a. heightening
b. flattening
c. trimming
d. peaking
e. circumventing

Answer: (b) Difficulty: (1)

68. The ________________ is the number of subordinates reporting directly to a supervisor.

a. authority level
b. chain of command
c. power channel
d. line-staff relationship
e. span of control

Answer: (e) Difficulty: (2)

69. If an organization structure is tall it generally has a ______________ span of control. (Pick the most appropriate term.)

a. dependent
b. narrow
c. wide
d. independent
e. counterproductive

Answer: (b) Difficulty: (2)

70. The way a company is organized—its structure—is a particularly important potential source of stress, but one that is manageable if the person doing the organizing knows what to look for. One of the most common stressors is ________________, where a situation exits in which two or more sets of demands are placed on an employee in such a manner that compliance with one demand makes it difficult to comply with the other.

a. role ambiguity
b. role empowerment
c. role conflict
d. role reduction
e. role re-alignment

Answer: (c) Difficulty: (3)

True/False - Short Answer

71. The scalar chain is arranging the activities of the enterprise in such a way that they systematically contribute to the enterprise's goals.

Answer: (False) Difficulty: (2)

72. Chain of command, scalar chain, and line of authority are terms that mean basically the same thing.

Answer: (True) Difficulty: (1) Page: 212

73. The ________________ of an organization shows the structure of the organization, specifically, the title of each manager's position and, by means of connecting lines, who is accountable to whom and who is in charge of what department.

Answer: (organization chart) Difficulty: (2)

74. Habitual contacts, communications, and ways of doing things that employees always develop is an illustration of the ________________ organization.

Answer: (informal) Difficulty: (3)

75. One thing the organization chart does not show is the informal organization that has evolved in the enterprise.

Answer: (True) Difficulty: (1)

76. Departmentalization is the process through which an enterprise's activities are grouped together and assigned to managers; it is the organization-wide division of work.

Answer: (True) Difficulty: (2)

77. Divisionalization is the process through which an enterprise's activities are grouped together and assigned to managers; it is the organization-wide division of work.

Answer: (False) Difficulty: (2)

78. Functional departmentalization means grouping activities around basic departments like manufacturing, sales, and finance.

Answer: (True) Difficulty: (1)

79. Service businesses differ from production-oriented businesses in that they are not built around business functions.

Answer: (False) Difficulty: (2)

80. _______________ means the firm's major departments are organized so that each can manage all the activities needed to develop, manufacture, and sell a particular product or product line.

Answer: (Divisionalization) Difficulty: (3)

81. An additional problem that small companies (such as Intuit—a software manufacturer) face is that divisionalization does not fit well with their organizational schemes because divisional units are more difficult to manage.

Answer: (False) Difficulty: (2)

82. ____________________ is similar to divisionalization except that departments are organized to serve the needs of specific customers.

Answer: (Customer departmentalization) Difficulty: (2)

83. The conduit (wholesaler, drugstore, grocery, or the like) through which a manufacturer distributes its products to its ultimate customers is called a company command structure.

Answer: (False) Difficulty: (2)

84. A _____________ organization is defined as the superimposition of one or more forms of departmentalization on top of an existing form.

Answer: (matrix) Difficulty: (2)

85. Dual reporting along product and geographic lines is a common matrix organization approach.

Answer: (True) Difficulty: (1)

86. Matrix organizations can trigger problems that can sometimes be serious. Among the most common of these problems are the initiation of power struggles and conflicts, lost time, and excessive overhead due to hiring more managerial and support talent.

Answer: (True) Difficulty: (1)

87. One of the techniques for achieving coordination is through mutual adjustment.

Answer: (True) Difficulty: (2)

88. ____________ is the right to take action, to make decisions, and to direct the work of others.

Answer: (Authority) Difficulty: (3)

89. Power in organizations has to be legitimate.

Answer: (False) Difficulty: (2)

90. The five common bases for power in an organization are reward power, coercive power, intelligence power, expert power, and espionage power.

Answer: (False) Difficulty: (2)

Essay

91. The text identifies five basic means to departmentalize an organization. First, list and briefly identify each of these five basic departmentalization forms. Second, using organizational diagrams of your own choosing, illustrate each form at the senior, director or vice-president, and manager level. These diagrams do not need to be complex, however, they do need to show the primary characteristic of the form.

Answer:

a. Functional departmentalization means grouping activities around basic functions like manufacturing, sales, and finance. Each department is organized around a different business function. Note: students can also describe departmentalization based on technology (even though it is not one of the primary forms).

b. Product departmentalization has departments for each of the company's products or for each family of products. Department heads in this type of organization are responsible for both producing and marketing a product or family of products.

c. Customer departmentalization is similar to divisionalization except that departments are organized to serve the needs of specific customers.

d. Marketing-channel departmentalization means that top-level departments are organized around each of the firm's marketing channels. In this method (as opposed to customer departmentalization), the same product is typically marketed through two or more channels. Usually, one department is chosen to manufacture the product for all the other marketing-channel departments.

e. Geographic or territorial departmentalization means that separate departments are organized for each of the territories in which the enterprise does business. Each geographic area or territory tends to be self-contained, perhaps with its own production, sales, and personnel activities.

Examples of drawings for the second phase of this question can be found in Figures 7.2 (functional departmentalization), 7.4 (product departmentalization), 7.5 (customer departmentalization), 7.6 (marketing-channel departmentalization), and 7.7 (geographic departmentalization).

Difficulty: (3)

92. Define or describe a matrix organization and briefly list the advantages and disadvantages.

Answer:

A matrix organization (also known as matrix management) is defined as the superimposition of one or more forms of departmentalization on top of an existing form.

Combinations can be created (most commonly) between products and functions or products and geography. Some of these organizations are more formal than others.

Advantages and disadvantages: Matrix departmentalization can help give bigger companies some of the advantages of smaller ones. In addition, management avoids having to establish duplicate functional departments for each of several projects.

However, matrix organizations can also trigger problems that are potentially serious. These problems can be listed as:

a. power struggles and conflicts.
b. lost time.
c. excessive overhead.

Difficulty: (2)

93. When work has been divided among units, it then needs to be coordinated. This is not always an easy task. Theorists Jay Galbraith and Henry Mintzberg have described several techniques that managers can use to achieve coordination. The text lists nine of these techniques. Pick any five, list them, and briefly describe their primary coordination function.

Answer:

a. Coordination through mutual adjustment. Mutual adjustment is used for coordination in both the simplest and most complex situations. This method relies on interpersonal communication (such as coordination to move a log).

b. Coordination through rules or procedures (standardization of work processes). If the work to be done is predictable and can be planned in advance, a supervisor can specify ahead of time what actions his or her subordinates should take. Thus, rules and procedures are useful for coordinating routine, recurring activities.

c. Coordination through direct supervision (using the hierarchy). Direct supervision achieves coordination by having one person coordinate the work of others, issuing instructions and monitoring their results.

d. Coordination through divisionalization. Organizing around divisions can facilitate coordination. The coordination is usually the responsibility of the senior managers.

e. Coordination through staff assistants. Some managers hire a staff assistant to make the job of coordinating subordinates easier. Advice and workload reduction are benefits of this method.

f. Coordination through liaisons. When the volume of contacts between two departments grows, some firms use special liaisons to facilitate coordination.

g. Coordination through committees. Many firms achieve coordination by appointing interdepartmental committees, task forces, or teams.

h. Coordination through independent integrators. An independent integrator is an individual or group that coordinates the activities of several interdependent departments. Integrators differ from liaison personnel in that integrators are independent of (not attached to) the departments they coordinate. Instead, they report to the manager who oversees those departments.

i. Coordination through standardized targets, skills, or shared values. Firms can also achieve coordination by standardizing the efforts of their employees in any of the above.

Difficulty: (3)

94. Power has been defined by Max Weber, a famous sociologist, as "the probability that one actor within a social relationship will be in a position to carry out his own will despite resistance, regardless of the basis on which the this probability rests." Building on this, social scientists have identified at least five bases or sources of power. List each form and provide an illustration of each.

Answer:

a. Reward power: providing positive rewards (such as money) or reducing negative rewards (such as poor working conditions)

b. Coercive power: group pressure, threats, or intimidation

c. Legitimate power: authority (as in a monarch) or the right to assign tasks

d. Referent power: identification with another (such as modeling after a famous athlete)

e. Expert power: often stems from a person's position in the organization; a gatekeeper; a budget officer

Difficulty: (2)

95. Explain the difference between delegation and empowerment.

Answer:

Delegation can be defined as being the pushing down of authority from supervisor to subordinate. Authority can be delegated, responsibility cannot. Empowerment (though intertwined with delegation) is a broader term since it means authorizing and enabling workers to their jobs. For example, the assembly workers at Toyota do not just have the authority to solve problems on the line. In addition, they are given the training, tools, and management support required to enable them to solve their problems. In this way, Toyota workers are empowered to continuously improve production quality.

Difficulty: (2)

Chapter 8
DESIGNING ORGANIZATIONS TO MANAGE CHANGE

INTRODUCTION

This chapter includes factors affecting how organizations are designed and structured, the initial redesign steps for making organizations more responsive, how managers can organize and lead team-based organizations, an analysis of network-based and boundaryless organizations, and a description of the horizontal corporation.

LECTURE OUTLINE

What's Ahead?

A Wave of Reorganizations

Factors Affecting Organizational Structure
- Organization and Environment: The Burns and Stalker Studies
- Organization and Technology: The Woodward Studies
- Synthesis: A Contingency Approach to Organizing

How Are Managers Moving Beyond Organic Organizations?
- Reduce Layers of Management
- Establish Mini-Units
- Reassign Support Staff
- Widen Spans of Control

Building Team-Based Structures
- Nature of Team-Based Organizations
- Designing Organizations to Support Teams

Network-Based Organizations
- Formal Networks
- Informal Networks
- Electronic Networking

Boundaryless Organizations
- What Are Boundaryless Organizations?
- How to Pierce the Four Organizational Boundaries
- The Horizontal Corporation

Summary

Case: How Skandia Group Looks into the Future
You Be the Consultant: Redesigning KnitMedia to Manage Change

ANNOTATED OUTLINE

What's Ahead?

The chapter begins with a discussion concerning how Hewlett-Packard recognized that its progressive management and technological breakthrough capabilities were being seriously eroded by too much "decision overhead" in its expanded levels of committees and hierarchy that had solidified into an unresponsive giant.

Notes:

A Wave of Reorganizations

A wave or reorganizations have swept across U.S. industry as managers have attempted to deal with the challenge of managing rapid change. Examples of organizational change efforts at ABB (Asea Brown Boveri), IBM, and Saturn are discussed. Many firms have been moving to self-managing work teams in a culture that emphasizes quality and teamwork.

Notes:

Factors Affecting Organizational Structure

Traditional organizational structures are being replaced by flatter, more responsive formats. Important factors driving the change efforts are the rate of change in the environment of the firm and the technology used by the firm to produce its work.

Notes:

Organization and Environment: The Burns and Stalker Studies. Burns and Stalker studied industrial firms to determine how the nature of a firm's environment affected the way the firm was organized and managed. They believed a stable, unchanging environment demanded a different type of organization than did a rapidly changing one. They studied the stable environment and the innovative environment. Burns and Stalker found that firms in an innovative environment needed to focus their organizational structure on responsiveness and creativity as opposed to efficiency in the stable environment.

The findings of Burns and Stalker led them to distinguish between mechanistic and organic organizations. Mechanistic organizations were characterized by close adherence to the chain of command, a functional division of work, highly specialized jobs, use of formal hierarchy for coordination, detailed job descriptions, interaction between employees on a vertical basis, and a tendency for behavior to be governed by instructions and decisions issued by supervisors. Organic organizations were characterized by little preoccupation with the chain of command, a more self-contained, divisionalized structure of work, job responsibility not viewed by employees as a limited field of rights, jobs that are not clearly defined in advance, a network of matrix structure of communication, lateral rather than vertical communication with emphasis upon consultation, and pervasive commitment to the tasks of the organization.

Burns and Stalker found that in mechanistic organizations lines of authority are clear, employees adhere to the chain of command, functional types of departmentalization prevail, each employee has a highly specialized job, the most important decisions are centralized, and the span of control is narrow. In organic organizations, employee jobs are always changing and lines of authority are not so clear, a product/divisional type of departmentalization prevails, job enlargement is prevalent, important decisions are made at lower levels, and spans of control are wider with more general supervision.

Notes:

Organization and Technology: The Woodward Studies. Woodward discovered that the production technology of a firm affects the way the firm should be organized. She classified companies as either unit and small-batch production, large-batch and mass production, or process production. She found that lines of authority and adherence to the chain of command are rigid in mass production firms but more informal and flexible in unit and process production firms. She found functional departmentalization in mass production firms and product type departmentalization in unit and process production firms. In mass production firms, jobs were highly specialized, but in unit and process production organizations, jobs were less specialized. Organizations tended to be centralized in mass production firms and decentralized in unit and process production companies. Unit and process production firms had narrower supervisory-level spans of control than mass production firms.

Notes:

Synthesis: A Contingency Approach to Organizing. The Burns and Stalker findings suggest that different organizational structures are appropriate for, or contingent on, different tasks. In the case of routine requirements, a highly structured organization is most efficient. In the case of unpredictable circumstances, creativity and entrepreneurial activities are encouraged by a flexible organizational structure.

Notes:

How Are Managers Moving Beyond Organic Organizations?

The requirement for survival in business today is the ability to rapidly adjust to unforseen and dramatic change. Management flexibility is achieved by flattening the levels of management, reorganizing around mini-units, decentralizing support staff, and widening spans of control.

Notes:

Reduce Layers of Management. In large, bureaucratic organizations – like traditional American railroads - several layers of middle managers were necessary to synthesize and communicate information (both ways) between top management to the operators in the field. It is possible, now that we have electronic data processing and communications devices, to reduce those middle management layers. The application of computer technology has made some downsizing possible.

Notes:

Establish Mini-Units. In some cases it has been possible to reorganize large companies into many small "Strategic Business Units." These small units can be operated efficiently because the employees know each other and can communicate more efficiently. The book doesn't mention the top management challenge of ensuring the efforts of these many small units are controlled and coordinated so that the entire system operates efficiently.

Notes:

Reassign Support Staff. This strategy has two possible benefits. A bloated central support staff can impede effective operations. In such a case they can be distributed to operating units and make a positive contribution to customer satisfaction.

Notes:

Widen Spans of Control. As employees are increasingly empowered to make decisions concerning their jobs, there is less need for close supervision. Managers' spans of control can widen significantly with no loss of organizational effectiveness. When Pratt and Whitney implemented such a strategy the organization ran leaner and customer service levels improved dramatically.

Notes:

Building Team-Based Structures

Nature of Team-Based Organizations. In team-based organizations, the team is the basic work unit. Teams are groups of people who work together and share a common work objective.

Notes:

Designing Organizations to Support Teams. Leaders in team-based organizations should function as coaches, use a participative style, and manage the interaction the team and the environment beyond its own boundaries. In order for a team-based organization to work effectively, management must have a philosophy emphasizing high employee involvement and trust, a flat organizational structure with few supervisors and delegation of decision making to teams, a reward system that supports the team-based approach, policies that support teamwork, and teams with a wide range of skills.

Notes:

Network-Based Organizations

Firms are superimposing organizational networks over their existing structures. An organizational network is a system of interconnected or cooperating individuals. The three types of networks are formal organizational, informal organizational, and electronic information.

Notes:

Formal Networks. Formal organizational networks are recognized groups of managers assembled by the CEO and the senior executive team. The number of managers involved is normally between 25 and 100. Networks are not temporary, they take the initiative to find and solve problems, and decentralize decision making for CEOs.

Notes:

Informal Networks. Informal networks consist of cooperating individuals who are interconnected only informally. They share information and work together to solve problems based upon their expertise

Notes:

Electronic Networking. Electronic networking consists of groups in an organization linked through the internet or collaborative computing.

Notes:

Boundaryless Organizations

Traditional organizations have clearly defined boundaries for authority, job specialization, lines of communication, and mechanical, step-by-step approaches to problem solving. Many organizations are piercing organizational boundaries to become more responsive.

Notes:

What Are Boundaryless Organizations? A boundaryless organization is one in which the widespread use of teams, networks, and other structural mechanisms mean that the boundaries that typically separate organizational functions and hierarchical levels are reduced and made more permeable. In a boundaryless company, internal organizational boundaries are stripped away as are the boundaries between the company and its suppliers and customers.

Notes:

How to Pierce the Four Organizational Boundaries. Four specific boundaries must be pierced if the company is to obtain full advantage of teams and networks. These include the authority boundary, the task boundary, the political boundary, and the identity boundary. To pierce the authority boundary, bosses must learn how to lead while remaining open to criticism and must be willing to accept orders from subordinates who are experts on a problem. Subordinates must be trained and encouraged to follow but still challenge superiors if an issue must be raised. Piercing the task boundary requires that employees forget the "It's not my job" attitude. To pierce the political boundary, managers and employees must ask "What's in it for us?" when decisions must be made, and not engage in turf battles with other departments. Employees tend to identify with the group with which they have a shared experience, and this may lead to an "us" versus "them" mentality. To break this down, employees must be educated about the overall goals of the organization and taught to avoid "devaluing the potential contribution of other groups."

Notes:

<u>**The Horizontal Corporation**</u>. The horizontal corporation is a structure organized around customer-oriented processes such as new-product development, sales and fulfillment, and customer support. Employees work together in multidisciplinary teams; each team performs one or more of the processes. The horizontal corporation usually has a small team of senior executives to ensure strategic direction and to provide essential staff support functions like human resource management. The essence of creating a horizontal corporation is defining the firm's core processes and then organizing teams around these while linking each process team's performance to specific customer-related objectives.

Notes:

Summary

Notes:

Case: How Skandia Group Looks into the Future

1. Other firms would be wise to implement Edvinsson's approach because it uses people in the environment who understand the archeology of the future and are likely to come up with a likely scenario for the future. Some firms have been reluctant to move away from doing strategic in-house planning at the corporate level. They would also be unlikely to use young 25-year-olds in the process, instead opting entirely for experienced strategic planners.

2. His approach would meet significant resistance in organizations with managers who are unlikely to readily accept change. These managers are accustomed to the old ways of doing strategic planning, and they are unlikely to accept a new approach unless they see a strong reason to change.

3. Edvinsson has redesigned Skandia with what appears to be a network approach. He has a corporate council made up of 150 of the firm's senior managers who will hear the report of the members of the Skandia Futures center. He is using teams with diverse functional roles, organizational experiences, and cultural backgrounds to improve performance and decision making.

You Be the Consultant: Redesigning KnitMedia to Manage Change

1. The organization chart should provide KnitMedia with the ability to manage rapid change because it is organized by product line and does not have a lot of hierarchy to go through for decisions. KnitMedia will need to clarify the roles of the CEO, COO, and CFO, however. They would probably function best as an executive committee.

2. KnitMedia could consider a team-based organization to help manage change. The teams could be self-contained and self-managing. Top management will have to be willing to delegate authority and continue emphasizing high involvement and trust. The organizational structure will have to be flat with

few supervisors and delegation of authority to the work teams. Company policies should support the team concept with equal treatment, egalitarian approaches to issues like parking access, and minimal status differences. Employment stability is important to help foster commitment. Teams must have a wide variety of skills to function well. These include job skills, interpersonal skills, team skills, and management skills. Networks could also be used with team members exchanging information and solving problems together electronically through the Internet. If the team-based approach is implemented successfully, boundaries that separate organizational functions and hierarchical levels can be reduced and made more permeable. These approaches in combination could be especially helpful for KnitMedia in responding more rapidly to change.

End of Chapter Questions for Chapter 8

Multiple Choice

1. At IBM, CEO Richard Gerstner, Jr., is stripping away the ponderous central staff and bureaucratic procedures that helped to slow IBM's responsiveness, substituting instead decentralized decision making and smaller organizational units. Which of the following terms best describes the process that IBM has just gone through?

a. Entrepreneurship
b. Intrapreneurship
c. Reorganization
d. Deconstruction
e. Realignment

Answer: (c) Difficulty: (2)

2. In the early 1990s, Zurich-based ABB (Asea Brown Boveri) cut its headquarters staff by 95 percent and moved 215,000 people into 5,000 largely independent "profit centers" (individual minicompanies) that average only 50 people each. Which of the following terms best describes the process that ABB (Asea Brown Boveri) has gone through?

a. Entrepreneurship
b. Intrapreneurship
c. Deorganization
d. Deconstruction
e. Realignment

Answer: (c) Difficulty: (3)

3. For a manager to understand why today's rapid change is causing traditional structures to be replaced by more responsive ones, it is useful to understand the factors that influence how companies should be structured. Two important factors to be understood are the rate of change of the firm's environment and (pick the answer that has the greatest impact on change according to the text) ____________________

a. the technology the firm uses to produce its work.
b. the changing labor force.
c. the changing international work place.
d. the changing legislation factors.
e. the changing demographics of the customers.

Answer: (a) Difficulty: (3)

4. For a manager to understand why today's rapid change is causing traditional structures to be replaced by more responsive ones, it is useful to understand the factors that influence how companies should be structured. Two important factors to be understood are the technology the firm uses to produce its work and (pick the answer that has the greatest impact on change according to the text) ____________________

a. the rate of change of the firm's environment.
b. the changing labor force.
c. the changing international work place.
d. the changing legislation factors.
e. the changing demographics of the customers.

Answer: (a) Difficulty: (3)

5. Researchers Tom Burns and G. M. Stalker studied about 20 industrial firms in the United Kingdom a number of years ago. Their purpose was to determine how the nature of a firm's environment affected the way the firm was organized and managed. After their studies, the researchers concluded that (pick the best answer) ____________________

a. there was no difference between management structures in varying environments.
b. a stable, unchanging environment demanded a different type of organization than did a rapidly changing one.
c. management structures were the result of tradition rather than change environments.
d. organization structure was derived from chaos.
e. organization structure, once derived, did not change frequently.

Answer: (b) Difficulty: (3)

6. According to research done by Burns and Stalker, a stable environment can be characterized as having all the following characteristics EXCEPT ____________________

a. Demand for the organization's product or service is stable and predictable.
b. There is an unchanging set of competitors.
c. Technological innovation and new product developments are evolutionary rather than revolutionary.
d. Government policies regarding regulation and taxation of the industry change little over time.
e. Management hierarchies are entrepreneurial and risk-takers.

Answer: (e) Difficulty: (2)

7. Researchers Tom Burns and G. M. Stalker studied about 20 industrial firms in the United Kingdom a number of years ago. Their purpose was to determine how the nature of a firm's environment affected the way the firm was organized and managed. One of their studies involved a rayon manufacturer. Research showed that to be successful in this industry, the parent firm had to keep costs down and be efficient as possible. Its existence, therefore, depended on keeping unexpected occurrences to a minimum, to facilitate maintaining steady, high-volume production runs. The above situation is characteristic of which of the following industry situations? (Choose the best general answer.) ________________

a. An innovative environment
b. A stable environment
c. A duopoloy
d. A monopoly
e. A competitive international environment

Answer: (b) Difficulty: (2)

8. According to research done by Burns and Stalker, an unstable environment can be characterized as having all the following characteristics EXCEPT ____________

a. Demand for the organization's product or service can change drastically.
b. There can be a sudden unexpected change in the nature of the organization's competition.
c. An extremely rapid rate of technological innovation and new product development is present.
d. Management hierarchies are adverse to taking risk and are not generally entrepreneurial in nature.
e. Quickly evolving government policies regarding regulation and taxation try to keep pace with the steam of new products being introduced.

Answer: (d) Difficulty: (2)

9. Consider the following industry examples: Internet software, electronics, and computers, the space industry, the deep sea industry, and the gene industry. All of the above industries are generally characterized as being ________________ environments.

a. innovative
b. stable
c. duopoloy
d. monopoly
e. cartel international

Answer: (a) Difficulty: (1)

10. Researchers Tom Burns and G. M. Stalker studied about 20 industrial firms in the United Kingdom a number of years ago. Their purpose was to determine how the nature of a firm's environment affected the way the firm was organized and managed. In one of their studies (the electronics industry), they found there was a deliberate attempt to avoid specifying individual tasks. Each worker's job might change daily as employees rushed to respond to the "problem of the day." The above situation would most likely fit which of the following environments? (Choose the best general answer given the information above.)

a. An innovative environment
b. A stable environment
c. A duopoloy
d. A monopoly
e. A cartel

Answer: (a) Difficulty: (2)

11. As one executive recently stated, "The first requirement of management is that it should make the fullest use of the capabilities of its members; any individual's job should be as little defined as possible, so that it will 'shape itself' to his special abilities and initiative." The above situation would most likely fit which of the following environments? (Choose the best general answer given the information above.)

a. An innovative environment
b. A stable environment
c. A duopoloy
d. A monopoly
e. A cartel

Answer: (a) Difficulty: (2)

12. Researchers Tom Burns and G. M. Stalker studied about 20 industrial firms in the United Kingdom a number of years ago. Their purpose was to determine how the nature of a firm's environment affected the way the firm was organized and managed. Their findings lead them to distinguish between two types of organizations. These types were called ____________

a. limited and full.
b. narrow and wide.
c. tall and flat.
d. mechanistic and organic.
e. simplistic and inorganic.

Answer: (d) Difficulty: (3)

13. According to research done by Burns and Stalker, ________________ organizations are characterized by close adherence to the chain of command, a functional division of work, highly specialized jobs, and the use of the formal hierarchy for coordination.

a. limited
b. narrow
c. flat
d. mechanistic
e. organic

Answer: (d) Difficulty: (2)

14. All of the following are characteristics of mechanistic organizations EXCEPT ____________

a. close adherence to the chain of command.
b. a more self-contained, divisionalized structure of work.
c. a functional division of work.
d. highly specialized jobs.
e. a tendency for interaction between employees to be vertical, "between superior and subordinate."

Answer: (b) Difficulty: (2)

15. According to research done by Burns and Stalker, ________________ organizations are characterized by little preoccupation with the chain of command, a network or matrix structure of communication, lateral rather than vertical communication, and jobs that are not clearly defined in advance.

a. full
b. wide
c. tall
d. mechanistic
e. organic

Answer: (e) Difficulty: (2)

16. All of the following are characteristics of organic organizations EXCEPT ________________

a. close adherence to the chain of command.
b. a more self-contained, divisionalized structure of work.
c. jobs that are not clearly defined in advance.
d. a network or matrix structure of communication.
e. lateral rather than vertical communication.

Answer: (a) Difficulty: (2)

17. British researcher Joan Woodward studied the way that certain forces affects the way a firm should be organized. According to her research, all of the following factors were found to have no real bearing on which organizational structure was chosen for an organization EXCEPT ______

a. a company's history.
b. a company's background.
c. a company's size.
d. a company's production technologies.
e. a company's policies and procedures.

Answer: (d) Difficulty: (2)

18. According to research done by British researcher Joan Woodward, which of the following factors was found to have the most impact on which organizational structure was chosen for an organization?

a. A company's history
b. A company's background
c. A company's size
d. A company's production technologies
e. A company's policies and procedures

Answer: (d) Difficulty: (2)

19. Using British researcher Joan Woodward's classification scheme for production technologies, a(n) ______ is characterized as being companies that produce one-at-a-time prototypes and specialized custom units to customer's requirements.

a. large-batch and mass production
b. unit and small-batch production
c. process production
d. innovation production
e. service production

Answer: (b) Difficulty: (2)

20. Using British researcher Joan Woodward's classification scheme for production technologies, a unit and small-batch production would best be characterized by which of the following examples?

a. A Cartier watch
b. Airplane manufacture
c. Ford cars
d. Oil refining
e. Disease screening in a public school

Answer: (a) Difficulty: (2)

21. Using British researcher Joan Woodward's classification scheme for production technologies, a(n) ___________ is characterized as being companies that produced products on assembly lines.

a. large-batch and mass production
b. unit and small-batch production
c. process production
d. innovation production
e. service production

Answer: (a) Difficulty: (2)

22. Using British researcher Joan Woodward's classification scheme for production technologies, a large-batch and mass production would best be characterized by which of the following examples?

a. a Cartier watch
b. fine piano manufacture
c. Ford cars
d. oil refining
e. disease screening in a public school

Answer: (c) Difficulty: (1)

23. Using British researcher Joan Woodward's classification scheme for production technologies, a(n) ___________ is characterized as being companies that produced products through continuously running facilities.

a. large-batch and mass production
b. unit and small-batch production
c. process production
d. innovation production
e. service production

Answer: (c) Difficulty: (2)

24. Using British researcher Joan Woodward's classification scheme for production technologies, a process production would best be characterized by which of the following examples?

a. A Cartier watch
b. Fine piano manufacture
c. Ford cars
d. Oil refining
e. Disease screening in a public school

Answer: (d) Difficulty: (2)

25. According to the contingency approach to organizing, a _____________ organization would be characterized as being one that had a stable environment, a classical organization, close adherence to the chain of command, and functional departmentalization.

a. pyramid
b. clover-leaf
c. X-Bar
d. organic
e. mechanistic

Answer: (e) Difficulty: (2)

26. According to the contingency approach to organizing, a _____________ organization would be characterized as being one that had an innovative environment, a behavioral organization, a flexible approach to the adherence to the chain of command, and divisional departmentalization.

a. pyramid
b. clover-leaf
c. X-bar
d. organic
e. mechanistic

Answer: (d) Difficulty: (2)

27. Today, says management expert Tom Peters, success in the marketplace "is directly proportional to the knowledge that an organization can bring to bear, how fast it can bring that knowledge to bear, and the rate at which it accumulates knowledge." Which of the following statements MOST ACCURATELY summarizes this view?

a. "You get what you pay for!"
b. The inefficient rise to their own level of mediocrity."
c. "Companies must be organized to respond to rapid change, and to respond very quickly."
d. "The Dilbert Principle rules most organizations."
e. "If something can go bad, it usually will."

Answer: (c) Difficulty: (2)

28. _______________ means drastically reducing the size of a company's workforce.

a. Upending
b. Paralleling
c. Downsizing
d. Cross-cutting
e. Synthesizing

Answer: (c) Difficulty: (1)

29. As managers downsized and moved beyond organic organizational structures, they took their existing structures and did all of the following EXCEPT ________________

a. increasing centralization to ensure control of top management.
b. reducing the levels of management.
c. reorganizing around small mini-units.
d. reassigning support staff from headquarters to the divisions.
e. empowering workers.

Answer: (a) Difficulty: (2)

30. Perhaps the MOST widespread tactic to manage change is characterized as being to ____________

a. widen span of control.
b. innovate.
c. establish smaller mini-units.
d. reduce layers of management.
e. reassign support staff to where they are needed.

Answer: (d) Difficulty: (3)

31. When he took over the Union Pacific Railroad in 1987, former CEO Mike Walsh found an extremely bureaucratic and sluggish organization. The company had nine layers of management. Walsh followed the most widespread tactic to manage change to begin to fix the railroad and its operating structure. Which of the following terms or phrases identifies the strategy that Mr. Walsh followed first as his main strategy?

a. Widen span of control
b. Innovate
c. Establish smaller mini-units
d. Reduce layers of management
e. Reassign support staff to where they are needed

Answer: (d) Difficulty: (2)

32. At Intuit, the new CEO "deorganized" his employees into eight separate businesses, each with its own general manager and mission. Which of the following strategies did the CEO most likely follow to accomplish this "deorganization?"

a. Widen span of control
b. Innovate
c. Establish smaller mini-units
d. Reduce layers of management
e. Reassign support staff to where they are needed

Answer: (c) Difficulty: (3)

33. New management structures present some interesting questions. For example, if supervisors are reduced, who will be there to supervise? The answer, increasingly, is the employees themselves. At one time, employees had to consult a supervisor before taking some action (like approving a refund). Today, they are making the decision themselves. Which of the following terms best describes this new atmosphere?

a. Downsizing
b. Realignment
c. Control
d. Empowerment
e. Democracy

Answer: (d) Difficulty: (2)

34. Pratt and Whitney engine division of United Technologies Corporation was in trouble with the airline industry. In fact, the situation had gotten so bad that American Airlines and United Airlines threatened to never buy another Pratt and whitney engine. The solution was to ________________ employees. They did this by expanding the number of service representatives in the field by 70 percent and gave these field reps the authority to approve multimillion- dollar warranty replacements on the spot (instead of requiring the affected client to wait months for headquarters approval). Because of this action, turnaround has been rapid.

a. downsize
b. realign
c. control
d. empower
e. democratize

Answer: (d) Difficulty: (2)

35. Even though such steps as empowering employees, widening spans of control, and establishing mini-units has been effective, the manager's job is still not done. The contemporary trend in redesigning organizations is to use teams, networks, and: (choose the best answer) ________________

a. "boundaryless" structures.
b. pyramid structures.
c. closed communication loops.
d. behavioral think tanks.
e. organizational democracy.

Answer: (a) Difficulty: (3)

36. A ________________ is a group of people who work together and share a common work objective.

a. cabal
b. cartel
c. clique
d. team
e. sub-division

Answer: (d) Difficulty: (1)

37. ______________ are empowered to manage themselves and make fast, on-the-spot decisions.

a. Sub-divisions
b. Teams
c. Self-managing work teams
d. Response units
e. Cabals

Answer: (c) Difficulty: (2)

38. Consider the following duties: recruit, hire, evaluate, and fire (if necessary); formulate, then track and amend, its own budget; make capital investment proposals as needed; and handle quality control, inspections, subsequent troubleshooting, and problem solving. The above MOST ACCURATELY describes the functions and powers of which of the following employee groups?

a. Sub-divisions
b. Teams
c. Self-managing work teams
d. Response units
e. Cabals

Answer: (c) Difficulty: (2)

39. A Chesebrough-Ponds employee recently remarked, "Empowerment is exciting. If we see something that will affect the quality to our customers, we have the freedom to stop a process. They [management] trust us." The above is an illustration MOST LIKELY to be descriptive of which of the following group formats?

a. Human resource sub-division
b. Teams
c. Self-managing work teams
d. Response units
e. Purchasing division

Answer: (c) Difficulty: (2)

40. Companies have traditionally been organized with individuals, functions, or departments as their basic work units. This was demonstrated on the organizational chart by giving or showing separate boxes for each functional department, down to separate tasks for individual workers at the bottom of the chart. In ____________, the team is the basic work unit. Decision making, planning, and implementing processes are completed by these new groups. (Pick the best descriptive answer.)

a. communication-based organizations
b. team-based organizations
c. technology-based organizations
d. individual-based organizations
e. client-based organizations

Answer: (b) Difficulty: (1)

41. No manager can create a team-based organization without providing the supporting mechanisms that will allow the teams to flourish. All of the following are organizational support mechanisms that enable work teams to do their jobs and ensure proper functioning of team-based organizations EXCEPT ______________

a. organizational therapy.
b. organizational philosophy.
c. organizational structure.
d. reward systems.
e. organizational policies.

Answer: (a) Difficulty: (2)

42. Companies in which work teams flourish are characterized by ideas and core values that emphasize high employee involvement and trust. Which of the following organizational support mechanisms best describes this or would contain this characterization?

a. Employee skills
b. Organizational philosophy
c. Organizational structure
d. Reward systems
e. Organizational policies

Answer: (b) Difficulty: (2)

43. Team-based organizational structures are characterized (with respect to the organizational structure support mechanism) as having all of the following EXCEPT ______________

a. flat structure.
b. few supervisors.
c. all or most of the work organized.
d. low involvement of employees.
e. high delegation of jobs.

Answer: (d) Difficulty: (2)

44. One of the typical policies that supports a team-based approach is to use ______________ incentives. Under this concept, employees are paid a portion of the savings incurred by their hard work; however, the payment is made to the team rather than to the individual employee. The team then distributes the money.

a. profit sharing
b. work-load sharing
c. gainsharing
d. fixed fee
e. piece-work

Answer: (c) Difficulty: (2)

45. All of the following policies help to ensure and build close-knit, cohesive work teams EXCEPT ______________

a. reserved parking places.
b. minimal status differences in offices and dress at the work place.
c. employees' evaluations of supervisors.
d. supervisor evaluations of employees.
e. policies of employment stability.

Answer: (a) Difficulty: (3)

46. Teams are only as good as the individuals on them. In order for a work team to operate effectively and efficiently, individual members must have skills that will aid the team. All of the following are important skills that should be or must be possessed by team members EXCEPT ______________

a. the job skills to actually do the job (such as welding).
b. the interpersonal skills to work in a healthy manner with and in the team.
c. extra-sensory skills so individuals can anticipate one another.
d. team skills (such as problem solving).
e. management skills.

Answer: (c) Difficulty: (1)

47. In general, a(n) ______________________ is a system of interconnected or cooperating individuals.

a. organizational division
b. corporation
c. organizational network
d. dyad
e. corporate cell

Answer: (c) Difficulty: (3)

48. A ______________ is, in essence, superimposed over the existing organizational structure, thus enhancing the likelihood that the work of far-flung units can be carried out promptly and in a coordinated way if quick decisions on some matters must be made.

a. dyad
b. cell
c. network
d. diagram
e. corporate paste-up

Answer: (c) Difficulty: (2)

49. Whether formal or informal, organizational networks share the same basic idea ____________

a. to make arrangements as simple as possible between managerial levels.
b. to link managers from different areas into multidisciplinary teams which communicate with one another across normal organizational boundaries.
c. to find new ways to downsize the organization.
d. to find new ways to increase the control of the top management and divisional vice-presidents.
e. to create formless organizations that are easier to change and respond to change.

Answer: (b) Difficulty: (3)

50. A(n) ____________ organizational network has been defined as "a recognized group of managers assembled by the CEO and the senior executive team.....The members are drawn from across the company's functions, business units, and geography, and from different levels of the hierarchy."

a. formal
b. informal
c. matrix
d. parallel
e. electronic

Answer: (a) Difficulty: (2)

51. Formal networks differ from teams, cross-functional task forces, or ad hoc groups in all of the following ways EXCEPT ________________

a. Networks are not temporary.
b. Networks do not concern themselves with profit or profit projection.
c. Networks take the initiative in finding problems.
d. Networks take the initiative in solving problems.
e. The existence of the formal network changes—or should change—the nature of top management's job.

Answer: (b) Difficulty: (2)

52. At Conrail, 19 middle managers from various departments and levels constitute the firm's operating committee. The managers influence most of the firm's key operating decisions through this committee. They meet for several hours on Monday mornings to review and decide on various tactical issues and work on longer-term issues such as five-year business plans as well. The above would be an example of which of the following network formats?

a. Informal
b. Formal
c. Matrix
d. Parallel
e. Electronic

Answer: (b) Difficulty: (2)

53. A(n) ____________ organizational network consists of individuals who cooperate with other individuals who are loosely interconnected. They share information and help solve each other's problems based on their personal knowledge of each other's expertise.

a. informal
b. formal
c. matrix
d. parallel
e. electronic

Answer: (a) Difficulty: (1)

54. Each of the following is a way to help to build informal organizational networks and relationships EXCEPT ____________

a. Design and use management development programs.
b. Move managers from facility to facility.
c. Aid managers in building lasting relationships within the company and sub-units.
d. Allow socialization and personal relationships to develop.
e. Encourage dating and marriage within the company and sub-units to create a family atmosphere within the company and its subsidiaries.

Answer: (e) Difficulty: (1)

55. Within Shell Oil Company international mobility has created what one might call a "nervous system" that facilitates both corporate strategic control and the flow of information throughout the firm. Widespread transfers have created a superior degree of communication and mutual understanding between headquarters and subsidiaries and between subsidiaries and themselves, as well as a stronger identification with the corporate culture, without compromising the local subsidiary cultures. The above would be an example of which of the following types of network formats?

a. Parallel
b. Formal
c. Matrix
d. Informal
e. Electronic

Answer: (d) Difficulty: (2)

56. If a new sales manager in Latin America needs to get in to see a new client, the manager might call another sales manager in Zurich (who was previously stationed in Latin America) to see about any potential contacts in the prospective clients firm. This would be an example of which of the following types of network formats?

a. Parallel
b. Formal
c. Matrix
d. Informal
e. Electronic

Answer: (d) Difficulty: (2)

57. The rise of the Internet and of special "collaborative computing" software lets companies make better use of networks and encourages all employees throughout the firm to network better and more often. This form of network is described as being the ________________ network in an organization.

a. parallel
b. formal
c. matrix
d. informal
e. electronic

Answer: (e) Difficulty: (1)

58. Price Waterhouse employs 18,000 accountants. Using ____________ networking, an accountant in Dublin can use the Internet or special "collaborative software" to gain help solving a problem from a colleague half a world away. Group decision support systems allow employees—even those in different countries—to brainstorm ideas and work together on projects. (Pick the most specific answer.)

a. parallel
b. formal
c. matrix
d. informal
e. electronic

Answer: (e) Difficulty: (2)

59. Companies are replacing vertical hierarchies with horizontal networks; linking together traditional functions through interfunctional teams; and forming strategic alliances with suppliers, customers, and even competitors." This would be an example of creating ______________ organizations. (Choose the best response.)

a. bounded
b. clover-leaf
c. X-bar
d. boundaryless
e. synthetic

Answer: (d) Difficulty: (1)

60. A(n) ______________ organization is one in which widespread use of teams, networks, and similar structural mechanisms attempts to reduce policies and procedures that typically separate organizational functions and hierarchical levels.

a. bounded
b. clover-leaf
c. X-bar
d. boundaryless
e. synthetic

Answer: (d) Difficulty: (2)

61. In practice, four specific boundaries must be pierced if the company is to obtain full advantage of teams and networks. All of the following fall into that category EXCEPT ____________

a. authority boundary.
b. leadership boundary.
c. task boundary.
d. political boundary.
e. identity boundary.

Answer: (b) Difficulty: (3)

62. Superiors and subordinates—even those found in self-managing teams or formal networks—always meet at an ______________, in every company. A common question to be answered is "Who is in charge of what?"

a. authority boundary
b. leadership boundary
c. task boundary
d. political boundary
e. identity boundary

Answer: (a) Difficulty: (1)

63. To pierce the authority boundary, all of the following are required EXCEPT ________________

a. Bosses must learn how to lead while remaining open to criticism.
b. A boss must be willing to accept "orders" from lower-ranking employees who happen to be experts on the problems at hand.
c. "Subordinates" must be encouraged to violate chain of command if a problem warrants.
d. "Subordinates" must be trained to follow but still challenge superiors if an issue must be raised.
e. "Subordinates" must be encouraged to follow but still challenge superiors if an issue must be raised.

Answer: (c) Difficulty: (2)

64. Creating a boundaryless organization also requires managing the ____________, which means changing the way employees feel about who does what when employees from different departments must divide up their work.

a. authority boundary
b. leadership boundary
c. task boundary
d. political boundary
e. identity boundary

Answer: (c) Difficulty: (3)

65. Members of each special-interest group in a boundaryless firm may still ask "What's in it for us?" when a decision must be made. But they have to be encouraged to take a more collegial, consensus-oriented approach, defending their interests without undermining the best interests of the team, network, or organization. This illustration is most closely associated with which of the following boundaries?

a. Authority boundary
b. Leadership boundary
c. Task boundary
d. Political boundary
e. Identity boundary

Answer: (d) Difficulty: (2)

66. Consider the following example: Manufacturing typically has a strong interest in smoothing out the demand for its products and in making the firm's products as producible as possible. Sales, on the other hand, has an equally legitimate interest in maximizing sales (even if it means taking in a lot of custom or last-minute rush orders). The result of such opposing agendas in a traditional organization can be a conflict at the department's ________________

a. authority boundary.
b. leadership boundary.
c. task boundary.
d. political boundary.
e. identity boundary.

Answer: (d) Difficulty: (2)

67. The ________________ means that we tend to relate with those groups with which we have shared experiences and with which we believe we share fundamental values.

a. authority boundary
b. leadership boundary
c. task boundary
d. political boundary
e. identity boundary

Answer: (d) Difficulty: (2)

68. The problem at the ________________ arises because people tend to trust those with whom they identify but distrust others. Attitudes like these can undermine the free-flowing cooperation that responsive, networking, or team-based organizations require.

a. authority boundary
b. leadership boundary
c. task boundary
d. political boundary
e. identity boundary

Answer: (d) Difficulty: (2)

69. The ________________ is a structure organized around customer-oriented processes such as new-product development, sales and fulfillment, and customer support. Employees work together in multidisciplinary teams; each team performs one or more of the processes.

a. vertical corporation
b. pyramid corporation
c. horizontal corporation
d. parallel corporation
e. cross-service corporation

Answer: (c) Difficulty: (2)

70. All of the following are reasons that companies organize horizontally EXCEPT ________________

a. Downsizing does not change the way that work still has to be done; therefore, horizontal structures are necessary.
b. Horizontal structures help to obliterate organizational boundaries.
c. Horizontal structures tend to discourage fiefdoms.
d. Horizontal structures give more power to top management.
e. Horizontal structures lend themselves to team formulation.

Answer: (d) Difficulty: (3)

71. Researchers have found that all of the following cultivate the kind of collective responsibilities that are necessary to make a horizontal corporation work EXCEPT ________________

a. changing the leadership (even down to the departmental and supervisory level).
b. making responsibilities overlap.
c. basing rewards on unit performance.
d. changing the physical layout.
e. redesigning work procedures.

Answer: (a) Difficulty: (3)

True/False - Short Answer

72. A wave of deorganizations is sweeping industry as managers grapple with the challenge of managing rapid change.

Answer: (True) Difficulty: (1)

73. Two important factors that are instrumental to the change environment sweeping away traditional structures in business and industry are the rate of change of the firm's profit plans and the psychology the firm uses to produce its work from workers.

Answer: (False) Difficulty: (3)

74. According to researchers Tom Burns and G. M. Stalker, a stable, unchanging environment demands a different type of organization than does a rapidly changing one.

Answer: (True) Difficulty: (1)

75. A stable environment can be characterized as having technological innovation and new-product developments that are evolutionary rather than revolutionary.

Answer: (True) Difficulty: (2)

76. According to Burns and Stalker, traditional organization's were "pyramids of knowledge" in that top management made most of the decisions and communicated them downward.

Answer: (True) Difficulty: (2)

77. An innovative environment can be characterized as one where, though technology exists, there is basically an unchanging set of competitors.

Answer: (False) Difficulty: (2)

78. Examples of the innovative environment present within many industries today is the railroad, retail department store, and the health insurance industry.

Answer: (False) Difficulty: (2)

79. Mechanistic organizations are characterized as having a close adherence to the chain of command and highly specialized jobs.

Answer: (True) Difficulty: (1)

80. Organic organizations are characterized as having a preoccupation with the chain of command and an expansive, industry-wide divisionalized structure of work.

Answer: (False) Difficulty: (2)

81. The chief contribution of British researcher Joan Woodward in the field of organizational design and change was that she discovered that a firm's production technology affects the way the firm should be organized.

Answer: (True) Difficulty: (2)

82. ________________ production companies produce products such as paper and petroleum products through continuously running facilities.

Answer: (Process) Difficulty: (3)

83. The Burns and Stalker findings, and Woodward's suggest that different organizational structures are appropriate for, or contingent on, different tasks.

Answer: (True) Difficulty: (2)

84. _____________ means dramatically reducing the size of a company's workforce.

Answer: (Downsizing) Difficulty: (2)

85. Widening the span of control is perhaps the most widespread tactic used to manage change.

Answer: (False) Difficulty: (3)

86. Increasingly today, steps such as empowering, widening spans, or establishing mini-units are no longer enough to aid in the change of organizational structures. Organizations are becoming "boundaryless."

Answer: (True) Difficulty: (2)

87. Companies in which work teams flourish are characterized by a management philosophy and core values that emphasize a high degree of employee of control and structure so the teams know the rules or boundaries.

Answer: (False) Difficulty: (2)

88. The command-and-control approach is giving way to the support-and-coach approach in many organizations.

Answer: (True) Difficulty: (1)

89. A ______________ organized network has been defined as "a recognized group of managers assembled by the CEO and the senior executive team....The members are drawn from across the company's functions, business units, and geography, and from different levels of the hierarchy."

Answer: (formal) Difficulty: (3)

90. A ________________ organization is one in which the widespread use of teams, networks, and similar structural mechanisms means that the boundaries that typically separate organizational functions and hierarchical levels are reduced and made more permeable.

Answer: (boundaryless) Difficulty: (3)

91. Superiors and subordinates—even those found in self-managing teams or formal networks—always meet at a(n) ________________ boundary, in every company.

Answer: (authority) Difficulty: (3)

Essay

92. Researchers Tom Burns and G. M. Stalker studied about 20 industrial firms in the United Kingdom a number of years ago. Their purpose was to determine how the nature of a firm's environment affected the way the firm was organized and managed. They discovered there were differences between a stable environment and an innovative environment. Illustrate the differences found between the characteristics of the two environments.

Answer:

a. A stable environment can be characterized as follows:

1) Demand for the organization's product or service is stable and predictable.

2) There is an unchanging set of competitors.
3) Technological innovation and new-product developments are evolutionary rather than revolutionary, in that necessary product changes can be predicted well in advance and the required modifications made at a leisurely pace.
4) Government policies regarding regulation and taxation of the industry change little over time.

An innovative environment is characterized as being:

1) Demand for the organization's product or service can change drastically, sometimes overnight, as competitors introduce radically improved products.
2) Sudden, unexpected changes in the nature of the organization's competitors.
3) An extremely rapid rate of technological innovation and new-product development. Organizations in innovative environments usually rely heavily on research and development for their survival.
4) Quickly evolving government policies regarding regulation and taxation that try to keep pace with the stream of new and more technologically advanced products being introduced by firms.

Difficulty: (2)

93. Summarize the differences found between mechanistic and organic organizations in the research done by Burns and Stalker.

Answer:

a. <u>Lines of authority</u>: In mechanistic organizations, the lines of authority are clear and everyone closely adheres to the chain of command. In organic organizations, employees' jobs are always changing and the lines of authority are not so clear. Here there is less emphasis on sticking closely to the chain of command and much more emphasis on speaking directly with the person who might have an answer to the problem.

b. <u>Departmentalization</u>: In mechanistic organizations (with their emphasis on efficiency), functional types of departmentalization prevail. In organic organizations (where flexibility is the rule), a product/divisional type of departmentalization prevails.

c. <u>Degree of specialization of jobs</u>: In mechanistic organizations, each employee has a highly specialized job at which he or she is expected to become an expert. In organic organizations, "job enrichment" is the rule.

d. <u>Delegation and decentralization</u>: In mechanistic organizations, most important decisions are centralized. In organic ones, more important decisions are made at lower levels; they are more decentralized.

e. <u>Span of control</u>: The span of control is narrow in mechanistic organizations, and there is close supervision. Spans are wider in organic organizations, and supervision is more general.

Difficulty: (3)

94. British researcher Joan Woodward's contribution to the study of organizations lies in her discovery that a firm's production technology (the processes it uses to produce its products or services) affects the way the firm should be organized. Summarize her specific findings in terms of organizational structure.

Answer:

a. Lines of authority: The lines of authority and adherence to the chain of command are rigid in mass production firms but more informal and flexible in unit and process production firms.

b. Departmentalization: There is a functional departmentalization in mass production firms and a product type of departmentalization in unit and process production firms.

c. Degree of specialization of jobs: Jobs are highly specialized in mass production firms and less so in unit and process production firms.

d. Delegation and decentralization: Organizations tend to be centralized in mass production firms, and decentralized in unit and process production firms.

e. Span of control: Unit and process production firms have narrower supervisory-level spans of control than do mass production firms.

Difficulty: (2)

95. Define downsizing and indicate the four primary reactions that downsizing brings to an organization.

Answer:

a. Downsizing means dramatically reducing the size of a company's workforce.

b. Reactions include:

1) Reduced levels of management, thus flattening their companies' structure.
2) Reorganized around small mini-units.
3) Reassigned support staff from headquarters to the divisions, thus decentralizing decisions.
4) Further decentralizing by empowering workers.

Difficulty: (1)

96. In setting up a boundaryless organization, four boundaries must be overcome. List the boundaries, list the key question that must be answered in the boundary, and give an illustration of tensions that develop due to this boundary.

Answer:

a. Authority Boundary: Key Question—"Who is in charge of what?" Tension: How to lead but remain open to criticism. How to follow but still challenge superiors.

b. Task Boundary: Key Question—"Who does what?" Tension: How to depend on others you don't control. How to specialize yet understand other people's jobs.

c. Political Boundary: Key Question—"What's in it for us?" Tension: How to defend one's interests without undermining the organization. How to differentiate between win-win and win-lose situations.

d. Identity Boundary: Key Question— "Who is—and isn't—us?" Tension: How to feel pride without devaluing others. How to remain loyal without undermining outsiders.

Difficulty: (2)

Chapter 9
STAFFING AND HUMAN RESOURCE MANAGEMENT

INTRODUCTION

This chapter explains why human resource management is increasingly important today, it discusses the types of laws that affect a company's HR activities, it includes analysis of employee selection procedures, and it discusses appraising and compensating employees.

LECTURE OUTLINE

What's Ahead?

Human Resource Management as a Core Function
- HR's Role as a Strategic Partner

Human Resources Legal Framework
- Equal Employment Laws
- Affirmative Action Programs
- Sexual Harassment
- Occupational Safety & Health
- Labor-Management Relations
- Other Employment Law Issues

Staffing the Organization
- Job Analysis
- Personnel Planning
- Employee Recruiting

Selection and Placement Techniques
- Application Blanks
- Testing for Employee Selection
- Interviews
- Guidelines for Interviewees
- Other Selection Techniques

Orientation and Training
- Orientation and Training in Practice

Appraising and Compensating Employees
- Employee Appraisal
- Basic Forms of Compensation
- Employee Benefits
- Promotions, Terminations, and Discipline

Summary

Case: Will the Glass Ceiling Ever Be Shattered ?
You Be the Consultant: Implementing an HR Program at KnitMedia

ANNOTATED OUTLINE

What's Ahead?

The Society for Human Resource Management Web site contains a wealth of information about the field of human resource management. Ben & Jerry's could have located a search firm to help select the CEO or they could have read articles giving them insight on effective selection.

Human Resource Management as a Core Function

Human resource management (HRM) is the management function devoted to acquiring, training, appraising, and compensating employees. In effect, all managers are human resource managers. Human resource management plays a key role in molding a company's work-force into a motivated and committed team to manage change more effectively and achieve objectives.

Notes:

HR's Role as a Strategic Partner. Strategic human resource management has emerged as an important management concern because employees today are central to the achievement of a competitive advantage and to manage the change process effectively. Human resource management plays a key role in strategy formulation and execution. HR can help strategy formulation by supplying information regarding the internal strengths and weaknesses of the firm. The ability of managers to create a highly committed, competent, and customer-oriented workforce is crucial to effective execution of strategy.

Notes:

Human Resource's Legal Framework

Equal Employment Laws. An important discrimination law includes Title VII of the 1964 Civil Rights Act that bars discrimination because of race, color, religion, sex, or national origin. This law is enforced by the Equal Employment Opportunity Commission. The Equal Pay Act of 1963 requires equal pay for men and women performing similar work. The Pregnancy Discrimination Act of 1978 prohibits discrimination in employment against pregnant women. The Americans with Disabilities Act of 1990 requires employers to make reasonable accommodations for disabled employees at work. The main purpose of the equal employment laws is to ensure that everyone has an equal opportunity of getting a job or being promoted at work.

Notes:

Affirmative Action Programs. Equal employment opportunity aims to ensure equal treatment at work, but affirmative action requires the employer to make an extra effort to hire and promote those in a protected group. Affirmative Action includes taking specific actions designed to eliminate the present effects of past discrimination.

Notes:

<u>**Sexual Harassment**</u>. Sexual harassment is unwelcome sexual advances, requests for sexual favors, and other verbal or physical conduct of a sexual nature that occur under conditions including the following: when such conduct is made, either explicitly or implicitly, a term or condition of an individual's employment; when submission to or rejection of such conduct by an individual is used as the basis for employment decisions affecting the individual; or when such conduct has the purpose or effect of unreasonably interfering with an individual's performance or creating an intimidating, hostile, or offensive work environment. To deal with sexual harassment, managers should take complaints seriously, issue and inform employees about company policy on the matter, establish an internal complaint procedure, establish a management response system, begin management training to increase their awareness of the issues, and discipline managers and employees involved in sexual harassment.

Notes:

<u>**Occupational Safety and Health**</u>. The Occupational Safety and Health Act was passed by Congress "to assure so far as possible every working man and woman in the nation safe and healthful working conditions and to preserve our human resources." The Act sets standards that apply to almost all workers in the United States. Companies can take action to improve safety and health by reducing unsafe conditions that can lead to accidents, hiring safety-prone people, emphasizing safety, using training to improve safety, setting specific loss-control goals, formulating and enforcing safety rules, and conducting safety and health inspections regularly.

Notes:

<u>**Labor-Management Relations**</u>. Employees are permitted to organize unions under the laws of the United States. The Norris-LaGuardia Act put limitations on the rights of employers to obtain injunctions in labor disputes; the Wagner Act permitted employees to form unions, engage in collective bargaining, and engage in concerted activities. Wagner also set up the National Labor Relations Board and cited unfair labor practices that could be committed by employers. The Taft-Hartley Act attempted to restore a balance of power between labor and management and cited unfair labor practices that could be committed by unions. The Landrum-Griffin Act protected union members from unfair practices perpetrated by their unions.

Notes:

<u>**Other Employment Law Issues**</u>. Other important labor laws include the Fair Labor Standards Act that specifies the minimum wage, overtime pay rules, and child labor law. The Employee Polygraph Protection Act outlaws almost all uses of the polygraph machine for employment purposes. Privacy laws may provide legal rights for employees in certain jurisdictions regarding who has access to information about their work history and job performance. Under the Whistleblower Protection Act, some employees who blow the whistle on their employers for the purpose of publicizing dangerous employer practices are entitled to legal protection.

Notes:

Staffing the Organization

Staffing refers to the filling of a firm's open positions. It is a process that includes job analysis, personnel planning, recruiting, interviewing, testing and selection, and training and development.

Notes:

Job Analysis. Job analysis is a procedure used to determine the duties of the jobs and the kinds of people who should be hired for them. Data acquired is used to develop job descriptions and job specifications.

Notes:

Personnel Planning. Personnel planning consists of forecasting personnel requirements, forecasting the supply of outside and internal candidates, and producing plans that describe how candidates will be hired, trained, and prepared for the jobs that will be opening up.

Notes:

Employee Recruiting. Employee recruiting includes attracting a pool of viable candidates. Sources of applicants include internal avenues, advertising, employment agencies, executive recruiters, referrals or walk-ins, older workers, and the Internet.

Notes:

Selection and Placement Techniques

Application Blanks. An application blank is a form that requests information about factors like education, work history, and hobbies. It is a means of collecting information for verification.

Notes:

Testing for Employee Selection. A test is a sample of a person's behavior. In HR, it is used to predict a person's success on the job. About two-thirds of employers are now using skill tests, e.g., typing, and only about 17 percent use personality tests. Management assessment centers include observations and appraisals of performance under realistic situations in the center. These centers can be used to evaluate applicants or current employees.

Research indicates that people have a "set point" for happiness, a genetically determined happiness level to which the person quickly gravitates, virtually no matter what failures or successes the person experiences. The studies suggest that the people who are the happiest today will also be the happiest ten years from now.

Notes:

Interviews. Managers can boost the reliability and validity of selection interviews by planning the interview, establishing rapport, closing the interview with time for questions, and reviewing the interview as soon as possible after its conclusion.

Notes:

Guidelines for Interviewees. Applicants may improve their prospects for success in interviews if they prepare well, uncover and relate themselves to the interviewer's real needs, think carefully before answering, pay attention to their appearance and enthusiasm, make a good first impression, and watch their nonverbal behavior.

Notes:

Other Selection Techniques. Reference checking is important to verify information supplied by the candidate, but it is difficult to obtain information because of privacy laws and employer concerns about defamation lawsuits. Honesty testing has proliferated, especially since the restrictions were placed on use of polygraph tests. Health exams are important because they can identify health problems that can increase absenteeism and accidents as well as detecting diseases that may be unknown to the applicant.

Notes:

Orientation and Training

Once employees are selected, they must be prepared to do their jobs. This is the purpose of orientation and training. Orientation means providing new employees with basic information about the employer. Training programs are used to ensure that the new employee has the basic knowledge required to perform the job satisfactorily.

Notes:

Orientation and Training in Practice. Orientation and training are important components in the process of developing a committed and flexible high-potential workforce and of socializing new employees. Mazda's Flat Rock Plant is again used to show an example. Evaluators at Mazda want to know if employees will stand behind the company and take the initiative to solve problems. New employees receive training in new problem-solving and analysis skills as well as how to read balance sheets. Mazda believes the orientation and training programs enable the firm to manage change more effectively.

Notes:

Appraising and Compensating Employees

Employee Appraisal. Employee performance is appraised regularly in most organizations to provide feedback to employees and provide an opportunity to correct performance deficiencies. Some firms are moving to a blank rating form where employees and their bosses list specific objectives to be accomplished during the appraisal period. These objectives may include training and development goals, teamwork goals, problem-solving, taking on leadership responsibility, and plans for personal contribution to team goals such as quality improvement. Basic forms of compensation include fixed salary or hourly wage, financial incentives, and gainsharing plans.

Notes:

Basic Forms of Compensation. Employee compensation refers to all work-related payments. Under the hourly wage system, employees are paid a fixed amount for each hour they work. This system is generally used for lower-skilled occupations. Salaried employees receive a fixed sum per week or month, no matter how any hours they work. Most professional positions are salaried; the reality is that these job-holders typically work in excess of a "minimum" forty-hour week. Some occupations are compensated through incentive-pay programs. Salespeople typically receive a "commission" based upon the quantity of goods they sell. Some sales compensation plans contain elements of both a salary and commission. Production workers' pay may be based upon some combination of hourly wages and an incentive for each "piece" they make. Some employees are offered merit awards as a reward for sustained superior performance.

Notes:

Employee Benefits. Employee benefits are supplements to wages or pay that employees get based on working for the organizations. Some benefits such as unemployment insurance and worker's compensation are legally mandated. Benefits can be used to build employee loyalty to the organization.

Notes:

Promotions, Terminations, and Discipline. Promotion generally means rewarding the employee's efforts by moving that person to a job with increased authority and responsibility. Because of downsizing and flattening in organizations today, many firms have had to rely on lateral promotions to broaden employees' experiences and to help them gain additional skills. A grievance is a complaint that an employee lodges against an employer regarding economic or noneconomic issues. Discipline without punishment is a disciplinary technique where employees are given an oral reminder for breaking a rule and a written reminder if it is broken again. Then a paid one-day decision making leave is mandated if another incident occurs in the next few weeks. If the rule is broken again, then the employee may be dismissed. Dismissals of employees should occur only on the basis of just cause and only after all reasonable steps to rehabilitate the employee have failed. There are also occasions where immediate dismissal is required, e.g., gross insubordination.

Notes:

Summary

Notes:

Case: Will the Glass Ceiling Ever Be Shattered?

1. Many in society still believe that women are the primary care-givers in the family and that they will have to interrupt their careers for family reasons. Whether this is true or not for an individual female manager, this perception makes it difficult to move up in the organization. Additionally, there are other barriers such as women being perceived as too emotional to make the difficult decisions required of a president or CEO.

2. Affirmative action programs with goals, not quotas, should be used to help advance women in organizations. The affirmative action plan should be explained to employees so that they understand how

it works and why it is being implemented. Women who seek to advance to upper echelons should continue working hard to gain an excellent performance record to improve their qualifications for promotion. Each woman needs to do a self-assessment and determine her career goals. It should not be assumed that all female managers (similar to all male managers) aspire to be in top management.

3. The glass ceiling is important for human resources because female employees who aspire to be in top management but who see no opportunity to make it are unlikely to be as motivated as they otherwise would be. An argument could be made also that there is a glass ceiling for the disabled and racial minorities, assuming the statistics are similar to those of females. As the workforce becomes more diverse and people see how managing diversity can contribute to organizational effectiveness, many of the current barriers are likely to be reduced.

You Be the Consultant: Implementing an HR Program at KnitMedia

1. To continue growing and developing new business, KnitMedia needs to be concerned with employee selection and development as well as compensation. The firm's approach in compensation should be tied to its corporate strategy. In selecting managers, the firm should consider using an assessment center to determine the potential of the candidate as the firm grows. The firm appears to have no one with this type of experience in house, so it should consider acquiring a contingency human resource manager until the firm reaches the size where a permanent manager is needed.

2. The human resource function needs to be implemented at headquarters but decentralized because the firm is operating in a number of countries, and it needs to know the laws and cultural issues that could effect company operations. A contingency human resource manager could be retained to develop and implement necessary HR programs and policies until the firm grows to the extent where a full-time human resources manager could be justified.

End of Chapter Questions for Chapter 9

Multiple Choice

1. We know that the organization chart only supplies the skeleton of the organization. Why is it so important to select the right person for the job? Pick the best response available.

a. Employees generate capital for the firm.
b. The right employees bring life to the company.
c. Choosing the right employees means you avoid law suits.
d. Choosing the right employees keeps them from the competition.
e. Getting the right employees is primarily luck.

Answer: (b) Difficulty: (1)

2. ______________ is the management function devoted to acquiring, training, appraising, and compensating employees.

a. Staff management
b. Supervisory management
c. Public relations management
d. Human resource management
e. Industrial relations management

Answer: (d) Difficulty: (2)

3. Which of the following terms is sometimes used to describe the human resource management function in a firm? (Choose the best answer available.)

a. Public relations
b. Labor
c. People
d. Human skills
e. Staffing

Answer: (e) Difficulty: (3)

4. All of the following are terms that are sometimes used to describe the human resource management function in a firm <u>EXCEPT</u> (pick the least likely) ____________

a. HRM.
b. personnel management.
c. psychology management.
d. staffing.
e. human skills and abilities management.

Answer: (c) Difficulty: (2)

5. The human resource management function is devoted to all of the following functions that pertain to the employees of a firm <u>EXCEPT</u> ____________

a. acquiring employees.
b. training employees.
c. appraising employees.
d. setting sales quotas for employees.
e. compensating employees.

Answer: (d) Difficulty: (2)

6. All of the following actions boost the need for motivated, self-directed, and committed employees EXCEPT ______________

a. Companies are flattening their pyramids.
b. Companies are relying more on self-managing work teams.
c. Companies are more likely to pay workers by the straight salary system.
d. Companies are getting closer to customers by empowering employees.
e. Companies are getting closer to customers by giving employees the authority to handle customer complaints and requests.

Answer: (c) Difficulty: (3)

7. Experts believe that ____________________ is now playing a central role in molding a company's workforce into a motivated and committed team. This workforce can now help the company and its management manage change more effectively. (Pick the best answer.)

a. human resource management
b. production management
c. line management
d. the staff selection and promotion process
e. visionary process

Answer: (a) Difficulty: (3)

8. ____________________ has been defined as "the linking of HRM with strategic goals and objectives in order to improve business performance and develop organizational cultures that foster innovation and flexibility."

a. Supervisory management
b. Staffing
c. Personnel management
d. Strategic human resource management
e. Strategic visioning

Answer: (d) Difficulty: (1)

9. ____________________ means accepting the human resource function as a strategic partner in formulating the company's strategies, as well as in executing those strategies through HR activities like recruiting, selecting, training, and rewarding personnel.

a. Supervisory management
b. Staffing
c. Personnel management
d. Strategic human resource management
e. Strategic visioning

Answer: (d) Difficulty: (1)

10. All of the following are illustrations of how human resources or human resource management may be useful in formulating strategy EXCEPT ____________

a. decisions on mergers or acquisitions dependent on existing labor relations and cultural fit.
b. the ability to supply competitive intelligence (such as details on incentive plans being used by competitors).
c. the ability to forecast sales based on knowledge of the consumer and his or her habits.
d. supplies information on the firm's internal strengths.
e. supplies information on the firm's internal weaknesses.

Answer: (c) Difficulty: (2)

11. It has been established that human resource management may be useful in formulating strategy for the firm. An example would be when IBM's decision to buy Lotus was promoted in part by IBM's conclusion that its own human resources were inadequate to enable the firm to reposition itself as an industry leader in networking systems, or at least to do so quickly enough. Which of the following areas would BEST describe how the HR function helped IBM?

a. Decisions on mergers or acquisitions dependent on existing labor relations and cultural fit
b. The ability to supply competitive intelligence (such as details on incentive plans being used by competitors)
c. The ability to forecast sales based on knowledge of the consumer and his or her habits
d. HR supplying information on the firm's internal strengths and weaknesses
e. HR monitoring compensation plans

Answer: (d) Difficulty: (3)

12. All of the following are ways that HRM plays a pivotal role in helping a firm to successfully execute and implement the firm's strategic plan EXCEPT ________________

a. HRM's aid in developing processes and its ability to create a highly committed and competent workforce.
b. HRM's ability to help set pricing policy that is a reflection of consumer demand.
c. HRM's involvement in a firm's downsizing efforts.
d. HRM's involvement in a firm's restructuring efforts.
e. HRM's role in diversity issues that help the firm deal with the global environment and its challenges.

Answer: (b) Difficulty: (2)

13. More than any other management function, ____________ is subject to the constraints of numerous federal, state, and local laws.

a. personnel
b. production
c. purchasing
d. auditing
e. research and development

Answer: (a) Difficulty: (1)

14. Among the most important of all laws affecting the personnel function in a firm are those which concern (pick the best general answer) ________________

a. recruitment and hiring.
b. equal employment laws prohibiting employment discrimination.
c. reciprocity.
d. role exclusion.
e. equal salaries between labor and management.

Answer: (b) Difficulty: (2)

15. Title VII of the ____________________ bars discrimination because of race, color, religion, sex, or national origin.

a. Vocational Rehabilitation Act of 1973
b. Civil Rights Act of 1991
c. Pregnancy Discrimination Act of 1978
d. Civil Rights Act of 1964
e. Human Relations Act of 1968

Answer: (d) Difficulty: (2)

16. All of the following were important equal employment laws enacted to foster equality in the work place EXCEPT the ________________

a. Vocational Rehabilitation Act of 1973.
b. Equal Pay Act of 1963.
c. Pregnancy Discrimination Act of 1978.
d. Civil Rights Act of 1964
e. Human Relations Act of 1968

Answer: (e) Difficulty: (3)

17. All of the following would be general guidelines to follow if an employer wishes to be in compliance with equal employment law EXCEPT _____________

a. Clearly place age restrictions in beginning of the employment ad.
b. Do not rely solely on word of mouth to publicize job opportunities.
c. Do not place "Help Wanted—Male" or "Help Wanted—Female" ads.
d. Do not ask about arrest records (to automatically omit the person from the job search).
e. Do not ask about religious preferences or national origin.

Answer: (a) Difficulty: (1)

18. Which of the following statements BEST describes what Title VII of the 1964 Civil Rights Act says?

a. Requires equal pay for men and women for performing similar work
b. Bars discrimination because of race, color, religion, sex, or national origin; instituted the EEOC
c. Ruled that job requirements must be related to job success
d. Prohibits discriminating against a person 40 or over in any area of employment because of age
e. Requires affirmative action in employment for veterans of the Vietnam War era

Answer: (b) Difficulty: (2)

19. Which of the following statements BEST describes what the Civil Rights Act of 1991 says?

a. Requires equal pay for men and women for performing similar work
b. Bars discrimination because of race, color, religion, sex, or national origin; instituted the EEOC
c. Ruled that job requirements must be related to job success
d. Prohibits discriminating against a person 40 or over in any area of employment because of age
e. Reverses *Wards Cove, Patterson*, and *Morton* decisions; places burden of proof back on employer and permits compensatory and punitive money damages for discrimination

Answer: (e) Difficulty: (3)

20. Whereas equal employment opportunity aims to ensure equal treatment at work, ________________ requires the employer to make an extra effort to hire and promote those in a protected group (women or minorities).

a. anti-age discrimination
b. legislative action
c. human relations protection
d. affirmative action
e. vocational rehabilitation

Answer: (d) Difficulty: (2)

21. Which of the following goals set out by a firm would be examples of trying to meet affirmative action guidelines?

a. Hire only the most qualified.
b. Promote more minorities to middle management jobs.
c. Always consider strength when hiring.
d. Hire the youngest and brightest for positions.
e. Always consider family educational history. Genetics are important.

Answer: (b) Difficulty: (1)

22. All of the following illustrate typical situations where sexual harassment may be considered as being a problem EXCEPT: (pick the least likely situation) ____________

a. insisting on sexual favors for some reward.
b. unwelcome sexual advances.
c. preference to not participate in alternative sexual lifestyles.
d. requests for sexual favors.
e. verbal or physical conduct of a sexual nature that occurs under conditions that affect employment.

Answer: (c) Difficulty: (3)

23. Managers should keep two facts in mind concerning sexual harassment. First, in addition to being unfair and detestable, it is also illegal. Second, it ____________

a. is primarily found only in manufacturing and government service.
b. is always directed toward females.
c. occurs primarily on the East and West Coasts.
d. is a widely occurring problem.
e. only began in the 1970s with the influx of new female employees to the labor force.

Answer: (d) Difficulty: (2)

24. There are a number of strategies (steps) that employers can take to avoid or deal with sexual harassment when it occurs. All of the following are pertinent EXCEPT ____________

a. Take all complaints seriously.
b. Issue a strong policy statement condemning such behavior.
c. Inform all employees about the policy prohibiting sexual harassment and of their rights under the policy.
d. In most instances, consider the accused to be considered guilty until proven innocent because of the sensitive nature of the offense to the accuser.
e. Discipline managers and employees involved in sexual harassment.

Answer: (d) Difficulty: (3)

25. There are a number of strategies (steps) that employees can take to avoid or deal with sexual harassment when it occurs. All of the following are pertinent EXCEPT (pick the least likely legal recourse) ____________

a. Make a verbal request to the harasser and the harasser's boss that the unwanted overtures cease because the conduct is unwelcome.
b. Try to get the harasser to agree to get psychological counseling for their problem.
c. Write a letter to the accused providing a detailed statement of the facts and a statement that the person wants the harassing activities to end immediately.
d. Report the unwelcome conduct and unsuccessful efforts to get it to stop to the harasser's manager or to the HR director (or both) verbally and in writing.
e. Consult an attorney about suing the harasser and possibly the employer.

Answer: (b) Difficulty: (2)

26. Perceptions of sexual harassment are often gender based and differ among employees. All of the following findings came from a recent federal study concerning sexual harassment EXCEPT (pick the statement that appears to be least true) ________________

a. Females are more likely to report sexual harassment than males.
b. Females are more likely to define more socio-sexual behaviors as sexual harassment than males.
c. Males are rarely sexually harassed.
d. Males are less likely to attribute responsibility for sexual harassment to the alleged harasser than are females.
e. Males are more likely to place blame for harassment on the female target than are females.

Answer: (c) Difficulty: (2)

27. The ______________________ was passed by Congress "to assure so far as possible every working man and woman in the nation safe and healthful working conditions and to preserve our human resources."

a. Federal Health Guidelines Act
b. Occupational Safety and Health Act
c. Safety Compliance Act
d. Health and Medical Standards Act
e. Physical and Mental Health Act

Answer: (b) Difficulty: (3)

28. All of the following are steps that companies can take to improve the safety and health of their workforces EXCEPT (pick the least likely response) ________________

a. hiring secret enforcement employees that will report unsafe acts and those that commit them.
b. reducing unsafe conditions that can lead to accidents.
c. hiring safety-prone people.
d. emphasizing safety.
e. setting specific loss-control goals.

Answer: (a) Difficulty: (1) Pages 292, 294

29. Under the laws of the United States and many other countries, employees are permitted to organize into unions. In the United States, the ________________ Act guarantees each employee the right to bargain with employers for union benefits.

a. Wagner
b. Taft-Hartley
c. Norris-LaGuardia
d. Landrum-Griffin
e. Robinson-Patman

Answer: (c) Difficulty: (3)

30. The ______________ Act prohibits unfair labor practices by unions against employers (like refusing to bargain with the employer).

a. Wagner
b. Taft-Hartley
c. Norris-LaGuardia
d. Landrum-Griffin
e. Robinson-Patman

Answer: (b) Difficulty: (3)

31. Under the ____________________, some employees who blow the whistle on their employers for the purpose of publicizing dangerous employer practices are entitled to legal protection.

a. Fair Labor Standards Act
b. Employee Polygraph Protection Act of 1988
c. Federal Privacy Act of 1974
d. New York Personal Privacy Protection Act of 1985
e. Whistle-Blower Protection Act of 1989

Answer: (e) Difficulty: (1) Page: 294

32. The term ___________ is often used to refer to actually filling a firm's open positions.

a. hiring
b. staffing
c. line-placing
d. organization charting
e. employing

Answer: (b) Difficulty: (2)

33. All of the following are steps that are usually included in the staffing process EXCEPT __________

a. job analysis.
b. personnel planning.
c. recruiting.
d. dismissal.
e. interviewing.

Answer: (d) Difficulty: (1)

34. ______________ is the procedure used to determine the duties of the jobs and the kinds of people (in terms of skills and experience) who should be hired for them.

a. Job description
b. Job specification
c. Job analysis
d. Job questionnaire
e. Personnel planning

Answer: (c) Difficulty: (3)

35. A job description is (read each answer carefully and pick the most accurate) ___________

a. a list of skills and aptitudes sought in people hired for the job.
b. the procedure used to determine the duties of the jobs and the kinds of people (in terms of skills and experience) who should be hired for them.
c. a list of duties showing what the job entails.
d. a form for determining the functions and duties of a job.
e. forecasts of personnel requirements, forecasts of outside candidates for jobs, and producing hiring and training plans.

Answer: (c) Difficulty: (3)

36. A job specification is (read each answer carefully and pick the most accurate) ___________

a. a list of skills and aptitudes sought in people hired for the job.
b. the procedure used to determine the duties of the jobs and the kinds of people (in terms of skills and experience) who should be hired for them.
c. a list of duties showing what the job entails.
d. a form for determining the functions and duties of a job.
e. forecasts of personnel requirements, forecasts of outside candidates for jobs, and producing hiring and training plans.

Answer: (a) Difficulty: (3)

37. A job analysis questionnaire is (read each answer carefully and pick the most accurate) _______________

a. a list of skills and aptitudes sought in people hired for the job.
b. the procedure used to determine the duties of the jobs and the kinds of people (in terms of skills and experience) who should be hired for them.
c. a list of duties showing what the job entails.
d. a form for determining and describing the functions, conditions, and duties of a job.
e. forecasts of personnel requirements, forecasts of outside candidates for jobs, and producing hiring and training plans.

Answer: (d) Difficulty: (2)

38. All of the following are illustrations of trends that are leading some experts to believe companies will eventually become "jobless" and teams will be formed where the differences between "your job" and "my job" will blur and disappear. Find the EXCEPTION. (pick the least likely trend)

a. flattening of organizations.
b. establishing self-contained mini-units.
c. reengineering.
d. unionization.
e. boundaryless organizations.

Answer: (d) Difficulty: (2)

39. All of the following are major category headings that might appear on a job analysis questionnaire for developing job descriptions EXCEPT (pick the least likely heading) ______________

a. Summary of Duties.
b. Special Qualifications.
c. Gender.
d. Equipment.
e. Decision Making.

Answer: (c) Difficulty: (2)

40. Personnel planning consists of (read each answer carefully and pick the most accurate) ______________

a. a list of skills and aptitudes sought in people hired for the job.
b. the procedure used to determine the duties of the jobs and the kinds of people (in terms of skills and experience) who should be hired for them.
c. a list of duties showing what the job entails.
d. a form for determining and describing the functions, conditions, and duties of a job.
e. forecasts of personnel requirements, forecasts of outside and inside candidates for jobs, and producing hiring and training plans.

Answer: (e) Difficulty: (2)

41. ______________ is attracting a pool of viable job applicants.

a. Job analysis
b. Personnel planning
c. Recruiting
d. Staffing
e. Interviewing

Answer: (c) Difficulty: (1)

42. All of the following are considered viable sources of employee recruiting EXCEPT (select the least viable or likely) ________________

a. internal sources of candidates.
b. advertising.
c. employment agencies.
d. immediate family and relatives.
e. executive recruiters.

Answer: (d) Difficulty: (1)

43. Recruiting employees can come from a variety of sources. One of the easiest to use is turning inward (the internal candidates). Filling open positions with inside candidates has both benefits and drawbacks. Which of the following would be considered to be a prime benefit of hiring internally? (Pick the best reason from those available.)

a. Inside candidates have a tendency to maintain status quo and do not cause problems.
b. Creativity is always higher with internal candidates.
c. Inbreeding is beneficial.
d. Employees see that competence is rewarded and morale and performance may be enhanced.
e. Inside candidates are always willing to follow a new direction.

Answer: (d) Difficulty: (2)

44. The greatest disadvantage presented for the prospective employee with a private employment agency is that (select the greatest disadvantage) ________________

a. the agencies generally charge a fee for their services.
b. the agencies rarely find qualified applicants.
c. the agencies rarely know about "great" jobs.
d. employers are reluctant to use private agencies.
e. agencies only deal with low-level jobs.

Answer: (a) Difficulty: (2)

45. Once an employer has a pool of applicants, the employer can turn to the screening and selecting process. These processes use one or more of the following techniques EXCEPT (pick the least likely form of evaluation) ________________

a. application blanks.
b. interviews.
c. tests.
d. lie detector tests.
e. reference checks.

Answer: (d) Difficulty: (1)

46. A(n) ________________ is basically a sample of a person's behavior. It can be used to predict a person's success on the job.

a. application blank
b. test
c. resume
d. profile
e. reference

Answer: (b) Difficulty: (1)

47. Of the screening tests listed below, which one appears to be most popular according to research for evaluation of potential employees?

a. skills tests
b. personality tests
c. handwriting tests
d. intelligence tests
e. order ability tests

Answer: (a) Difficulty: (3)

48. In a(n) ___________________ , about a dozen management candidates spend two or three days performing realistic management tasks (like making presentations) under the observation of expert appraisers. Then each candidate's potential for management is thereby assessed.

a. profile center
b. management stress center
c. management training center
d. management assessment center
e. psychological profile and assessment center

Answer: (d) Difficulty: (3) Page: 303

49. A manager can boost the reliability and validity of selection interviews by following sound interviewing procedures. All of the following are considered to be sound procedures EXCEPT (pick the least applicable or sound) ____________________

a. planning the interview.
b. putting the applicant under stress in order to observe reactions.
c. establishing rapport.
d. asking questions.
e. always reviewing the interview once it is completed.

Answer: (b) Difficulty: (2)

50. Which of the following are best assessed through interviews rather than through some other screening device?

a. What are your skills?
b. What is your technical expertise?
c. What are you like as a person?
d. What do your past employers think of you?
e. What is your educational history?

Answer: (c) Difficulty: (3)

51. If a person wanted to get an extra edge in preparation for an interview all of the following would be good suggestions to follow <u>EXCEPT</u> (pick the least likely) ________________

a. Preparation is essential.
b. Talk as much as possible (literally on any subject).
c. Uncover the interviewer's real needs.
d. Relate yourself to the interviewer's needs.
e. Think before answering.

Answer: (b) Difficulty: (1)

52. Employee ______________ means providing new employees with basic information about the employer.

a. manuals
b. training programs
c. orientation
d. socializing
e. appraising

Answer: (c) Difficulty: (1)

53. Employee ____________________ are aimed at ensuring that all new employees have the basic knowledge required to satisfactorily perform the job.

a. manuals
b. training programs
c. orientation
d. socializing
e. appraising

Answer: (b) Difficulty: (2)

54. _______________ means transforming new employees into effective organizational members by steeping them in the values and traditions of the organization.

a. Training programs
b. Employee benefit analysis
c. Orientation
d. Socialization
e. Maturation

Answer: (d) Difficulty: (2)

55. Once employees have been at work for some time, their performance should be appraised, or evaluated. Which of the following is one of the purposes of appraisal? (Pick the best answer.)

a. Providing feedback on how employees are doing
b. Providing evidence for termination
c. Providing evidence for inclusion in downsizing efforts
d. Providing a means to strictly enforce company policies
e. Providing a means to separate loyal and disloyal employees

Answer: (a) Difficulty: (1)

56. _________________ refers to all work-related pay or rewards that go to employees.

a. Appraisal
b. Employee benefits
c. Employee compensation
d. Employee rights and entitlements
e. Labor relations

Answer: (c) Difficulty: (2)

57. If employees are compensated based on a set period of time (like a week, month, or a year), they are on a ________________ system.

a. hourly wage
b. commission
c. flexible
d. combination retainer plan
e. fixed salary

Answer: (e) Difficulty: (2)

58. A(n) __________________ is any financial reward that is contingent on a worker's performance.

a. gainsharing plan
b. financial incentive
c. compensation plan
d. benefits plan
e. retainer system

Answer: (b) Difficulty: (2)

59. A (n) ________________ is one where an incentive plan engages many or all of the employees in a common effort to achieve a company's productivity objectives; the resulting cost-savings gains are shared among employees and the company.

a. gainsharing plan
b. financial incentive
c. compensation plan
d. benefits plan
e. retainer system

Answer: (a) Difficulty: (1)

60. A (n) __________________ is any supplements to wages or pay that employees get based on working for the organization.

a. gainsharing plan
b. financial incentive
c. compensation plan
d. employee benefits plan
e. retainer system

Answer: (d) Difficulty: (3)

61. A(n) __________________ generally means rewarding the employee's efforts by moving that person to a job with increased authority and responsibility.

a. compensation
b. reward
c. gain
d. promotion
e. benefit

Answer: (d) Difficulty: (2)

62. A complaint that an employee lodges against an employer, usually regarding wages, hours, or some condition of employment, is called a ____________________

a. punishment.
b. gripe.
c. declaration.
d. consequence.
e. grievance.

Answer: (e) Difficulty: (1)

63. One way to avoid problems with grievances and employee unrest is to set up a FRACT model to use in disciplinary situations. All of the following are portions of this model EXCEPT ____________

a. firing.
b. reason.
c. audit.
d. consequences.
e. type.

Answer: (a) Difficulty: (3)

True/False - Short Answer

64. ________________ is the management function devoted to acquiring, training, appraising, and compensating employees.

Answer: (Human resource management) Difficulty: (3)

65. All managers are, in a sense, personnel managers.

Answer: (True) Difficulty: (1)

66. As workers become more fully empowered, the HR function becomes less important because the workers begin to manage themselves.

Answer: (False) Difficulty: (2)

67. The fact that employees are central to helping companies achieve competitive advantage and manage change has led to the emergence of strategic human resource management.

Answer: (True) Difficulty: (1)

68. ________________ is the "linking of HRM with strategic goals and objectives in order to improve business performance and develop organizational cultures that foster innovation and flexibility."

Answer: (Strategic human resource management) Difficulty: (3)

69. One of the problems that HR has faced for years is that it has lacked any significant role in formulating strategy. It continues in this capacity today.

Answer: (False) Difficulty: (2)

70. Small firms, unlike large firms, do not have to cover all the human resource functions (such as labor relations, safety, equal employment, and training). Therefore, management of the HR function is generally an easier task.

Answer: (False) Difficulty: (2)

71. One of the problems faced by the personnel function in a firm is that more than any other management function, personnel is subject to the constraints of numerous federal, state, and local laws.

Answer: (True) Difficulty: (1)

72. The equal employment laws prohibiting employment discrimination are among the most important personnel laws.

Answer: (True) Difficulty: (1)

73. The Equal Pay Act of 1963 bars discrimination because of race, color, religion, sex, or national origin. This act instituted the EEOC.

Answer: (False) Difficulty: (3)

74. Employers should confirm that the selection tests they use do not unfairly screen out minorities.

Answer: (True) Difficulty: (1)

75. Courts have ruled that it is all right to ask questions about an applicant's ethnic, racial, or marital status (as long as another person is in the room to verify that no improper sexual questions were asked).

Answer: (False) Difficulty: (2)

76. Sexual harassment has been interpreted as only meaning "the insistence of sexual favors in return for some reward."

Answer: (False) Difficulty: (2)

77. Based on various studies, it has been found that females are more likely than males to report that they experienced some form of unwelcome sexual attention and to define more social-sexual behaviors as sexual harassment than are males.

Answer: (True) Difficulty: (1)

78. The ________________ Act was passed by Congress "to assure so far as possible every working man and woman in the nation safe and healthful working conditions and to preserve our human resources."

Answer: (Occupational Safety and Health) Difficulty: (3)

79. The Norris-LaGuardia Act guarantees each employee the right to bargain with employers for union benefits.

Answer: (True) Difficulty: (2)

80. The Taft-Hartley Act protects union members from unfair practices perpetrated against them by their unions.

Answer: (False) Difficulty: (2)

81. The term ______________ is often used to refer to actually filling a firm's open positions.

Answer: (staffing) Difficulty: (2)

82. Job ________________ is the procedure used to determine the duties of the jobs and the kinds of people (in terms of skills and experience) who should be hired for them.

Answer: (analysis) Difficulty: (3)

83. According to information in the text, there is a trend toward companies becoming "jobless."

Answer: (True) Difficulty: (2)

84. When companies begin to search for recruits for positions that have come available in the firm, the largest source of these recruits is ______________________.

Answer: (current employees or internal sources) Difficulty: (3)

85. A test is basically a sample of a person's behavior.

Answer: (True) Difficulty: (1)

86. The American Psychological Association reports (with respect to honesty testing in the work place) that there is a "preponderance of evidence" that supports the notion that few (if any) tests really are able to accurately measure a person's honesty in the work place.

Answer: (False) Difficulty: (2)

Essay

87. The fact that employees today are central to helping companies achieve competitive advantage and manage change has led to the emergence of strategic human resource management. Define this term and indicate HR's role in formulating and executing strategy.

Answer:

a. Strategic human resource management has been defined as "the linking of HRM with strategic goals and objectives in order to improve business performance and develop organizational cultures that foster innovation and flexibility."

b. In many companies, HRM already plays a crucial role in formulating strategy. HR gives the firm an assessment of whether any new ventures, mergers, or acquisitions are a good cultural fit with the firm and whether changes would need to be made to avoid problems. In addition, HRM is in a unique position to supply competitive intelligence that may be useful in the strategic planning process. Details regarding advanced incentive plans being used by competitors, opinion surveys from employees that elicit information about customer complaints, and information about pending legislation like labor laws or mandatory health insurance are some examples.

c. HRM plays a pivotal role in successfully executing and implementing a company's strategic plan. HR can make sure that the right kind of worker is hired to fit a company's strategy. For example, if a company chooses to pursue a policy of customer satisfaction, then the company should be hiring committed, competent, and customer-oriented workers. Since HRM is involved in most restructures or downsizing, outplacement processes can make these difficult times more compatible for employees.

Difficulty: (2)

88. Explain what each of the following significant equal employment opportunity laws are and what is the main thrust of the legislation:

a. Title VII of the 1964 Civil Rights Act.
b. Equal Pay Act of 1963.
c. Vocational Rehabilitation Act of 1973.
d. Civil Rights Act of 1991.

Answer:

a. Title VII of the 1964 Civil Rights Act bars discrimination because of race, color, religion, sex, or national origin. Instituted the EEOC.

b. Equal Pay Act of 1963 requires equal pay for men and women for performing similar work.

c. Vocational Rehabilitation Act of 1973 requires affirmative action to employee and promote qualified disabled persons and prohibits discrimination against disabled persons.

d. Civil Rights Act of 1991 reverses *Wards Cove, Patterson,* and *Morton* decisions; places burden of proof back on employer and permits compensatory and punitive money damages for discrimination.

Difficulty: (3)

89. Define what you perceive to be sexual harassment. Next, list and briefly explain the steps that an employer might take to avoid or deal with sexual harassment when it occurs.

Answer:

Students should remember that sexual harassment is more than insisting on sexual favors in return for some reward. It also includes unwelcome sexual advances, requests for sexual favors, and other verbal or physical conduct of a sexual nature that occurs under conditions that makes another person feel uncomfortable.

Steps include: (1) Take all complaints about harassment seriously; (2) Issue a strong policy statement condemning such behavior; (3) Inform all employees about the policy prohibiting sexual harassment and of their rights under the policy; (4) Establish a complaint procedure so that all employees understand the chain of command when it comes to filing and appealing sexual harassment complaints; (5) Establish a management response system that includes an immediate reaction and investigation by senior management when charges of sexual harassment are made; (6) Begin management training sessions with supervisors and increase their own awareness of the issues; and, (7) Discipline managers and employees involved in sexual harassment.

Difficulty: (2) Pages 290, 291

90. List and briefly discuss the steps in the staffing process.

Answer:

There are six recognized steps in the staffing process. They are:

a. Job analysis: This is the procedure used to determine the duties of the jobs and the kinds of people (in terms of skills and experience) who should be hired for them.

b. Personnel planning: This step consists of three activities. First, forecasting personnel requirements (in terms of future open positions). Second, forecasting the supply of outside candidates and internal candidates. Third, producing plans that describe how candidates will be hired, trained, and prepared for the jobs that will be opening up.

c. Recruiting: This is attracting a pool of viable job applicants.

d. Interviewing: This is done through questioning, checking references, testing for honesty, requiring health exams, and determining aptitudes.

e. Testing and Selection: A test is basically a sample of a person's behavior. It is used to predict success in the work place.

f. Training and Development: Training begins with an orientation and continues throughout the employee's stay with the company.

Difficulty: (2)

Chapter 10
BEING A LEADER

INTRODUCTION

Students will study how to think and assess a situation like a leader, describe the foundations of leadership, discuss how to provide a vision for an organization, compare and contrast the various styles a leader can use, and list the specific attributes of being a leader.

LECTURE OUTLINE

What's Ahead?

The Challenge of Leading
- Leaders Fill Many Roles
- Leaders and Managers

Thinking Like a Leader
- Identify What Is Happening
- Account for What Is Happening
- Decide on the Leadership Actions You Will Take

What Are the Foundations of Leadership?
- Power
- The Leader's Traits

How Leaders Provide a Vision
- Setting a Direction: Visions, Missions, Objectives

The Leader's Style: How Leaders Act Like Leaders
- The Leader's Behavior
- Structuring and Considerate Styles
- Participative and Autocratic Leadership Styles
- A Model for Deciding When to Be Participative
- The University of Michigan Studies
- Transformational Leadership Behavior
- Are There Gender Differences in Leadership Style?

Being a Leader
- Thinking Like a Leader
- Develop Your Judgment
- Develop Your Other Leadership Traits
- Are You Ready to Be a Leader?
- Rating a Manager's Power
- Make Sure Your Followers Share Your Vision
- Adapt Your Leadership Style and Actions to the Situation
- Substitutes for Leadership: Use Your Other Management Skills to Help You Lead

Summary

Case: The Ultimate Entrepreneur
You Be the Consultant: Michael Dorf Leads KnitMedia

ANNOTATED OUTLINE

What's Ahead?

Barry Gibbons exercised leadership when he influenced the owners, employees, and bankers to work enthusiastically toward downsizing and diversifying the Spec's Music Stores.

The Challenge of Leading

Leadership means influencing others to work willingly toward achieving objectives. This chapter attempts to explain leadership theories and how they can be translated into leadership skills.

Notes:

Leaders Fill Many Roles. Effective leaders must understand more than just leadership theories--they must understand motivation, group interaction and behavior, culture, individual behavior, communication and organizational change.

Notes:

Leaders and Managers. Managers plan, organize, lead, and control so that leading and managing are inseparable in management theory. Leading is a part of managing--it is the behavioral and influencing part of managerial actions.

Notes:

Thinking Like a Leader

Leaders need critical leadership thinking skills that allow them to analyze a situation and discover the underlying assumptions and values that are motivating subordinates and to evaluate the evidence and think through how to apply what the leader knows about leadership theory to solve the problem. Leaders need to be able to identify what is happening, account for it, and decide on the necessary leadership actions.

Notes:

Identify What Is Happening. The first step in problem-solving is to define the problem.

Notes:

Account for What Is Happening. Use behavioral science principles to account for *why* the issues identified in the first step are occurring.

Notes:

Decide on the Leadership Actions You Will Take. What sort of leadership action will remedy the situation? There may be more than one correct answer.

Notes:

What Are the Foundations of Leadership?

Power. Leaders without power are really not leaders because they have no chance of influencing anyone to do anything. Leaders in organizations normally derive much of their power from their formal position and the ability to allocate rewards. In some cases, leaders may have expert or referent power depending upon their individual characteristics.

Notes:

The Leader's Traits. Researchers have studied the traits of successful leaders for many years in an effort to identify a set of core traits that would predict success as a leader. Recent research indicates that there are certain core traits that significantly contribute to success for a business leader. These include drive, the desire to lead, honesty/integrity, self-confidence, cognitive ability and knowledge of the business.

Notes:

How Leaders Provide a Vision

To be effective, leaders must provide a vision that is a general statement of the organization's intended direction that evokes positive emotional feelings in organization members.

Notes:

Setting a Direction: Visions, Missions, Objectives. Communicating a vision is especially important in the rapidly changing environment that describes the scenario for most organizations, today. The mission statement of the firm operationalizes the vision of top management. Objectives are specific results or outcomes that the leader wants the group to achieve.

Notes:

The Leader's Style: How Leaders Act Like Leaders

The Leader's Behavior. Leadership studies that focus on how the leader's style is related to his/her effectiveness as a leader all focus on what the leader does and how he/she behaves in trying to influence followers. These studies also focus on the two major functions of leaders – accomplishing the task and satisfying the needs of group members.

Notes:

Structuring and Considerate Styles. Initiating structure and consideration have been two of the most frequently used descriptions of leader behavior. These concepts evolved from the Ohio State University leadership studies. Initiating structure is leader behavior whereby the person organizes work to be done and defines relationships or roles, the channels of communication, and ways of getting jobs done. Consideration is leader behavior indicative of mutual trust, friendship, support, respect, and warmth. In most situations, considerate leaders will have more satisfied subordinates, but the effects of such considerate leadership on employee performance are inconsistent. The effects of initiating structure are also inconsistent with respect to performance and satisfaction.

Notes:

Participative and Autocratic Leadership Styles. Leaders can act in either a participative or autocratic style. Autocratic leaders solve problems and make decisions by themselves based upon information available at the time. Participative leaders share the problem with subordinates as a group, and together, they generate and evaluate alternatives and attempt to reach consensus on a solution.

Notes:

A Model for Deciding When to be Participative. Vroom, Jago, and Yetton have developed a technique that enables a leader to analyze a situation and decide whether it is appropriate for participation. The technique includes a set of management decision styles, a set of diagnostic questions, and a decision tree for identifying how much participation is called for in a situation.

Notes:

The University of Michigan Studies. Likert and his associates at Michigan identified two leadership styles. Employee-centered leaders focus on the individuality and personality needs of their employees and emphasize building good interpersonal relationships. Production or job-centered leaders focus on production and the technical aspects of the job. Likert concluded; "Supervisors with the best record of performance focus their primary attention on the human aspects of their subordinates' problems and on endeavoring to build effective work groups with high performance goals." The Michigan researchers also studied close and general leadership styles. Close supervision is usually associated with lower employee morale, and there is no consistent relationship between closeness of supervision and employee performance.

Notes:

Situational Theories of Leadership. A conclusion that could be drawn from the leadership theories is that the style of leadership right for one situation might not be appropriate for another. In Fiedler's Contingency Theory of Leadership, Fred Fiedler used a least preferred coworker (LPC) scale to measure approaches to leadership and to try and determine whether a leader who was lenient in evaluating associates was more likely or less likely to have a high-producing group than the leader who was demanding and discriminating. Three factors combine to determine which leadership style is more appropriate: position power or the degree to which the position enables the leader to get his/her group members to comply with and accept his/her decision and leadership, task structure, or how routine and predictable the work group's task is, and leader-member relations, or how well the leader gets along with workers and the extent to which they have confidence in and are loyal to him/her. Fiedler concluded that if the situation is favorable or unfavorable to the leader, a more task-oriented, low-LPC leader is appropriate. In the middle range where the factors are more mixed, a more people-oriented, high-LPC leader is more appropriate. Recent research findings cast doubt on the validity of these conclusions.

Notes:

The Path-Goal Leadership Theory is based upon expectancy theory which states whether a person will be motivated depends on whether the person believes he/she has the ability to accomplish a task and his/her desire to do so. The path-goal theory, developed by House, concludes that leaders should increase the personal rewards subordinates receive for attaining goals and make the path to these goals easier to follow. The leadership style required depends upon the situation, so the leader must be flexible and adopt the style that is required.

Notes:

The Leader-Member Exchange Theory (LMX) says that leaders may use different styles with different members of the same work group. Followers tend to fall in either the in-group or the out-group in relating to the leader. The quality of leader-member exchanges was positively related to a leader's perception of follower attitudes and extroversion. Followers whose attitudes were similar to the leader and who were more extroverted were more likely to have high-quality leader-member exchanges. Leaders should try to make the in-group more inclusive and followers should try to be in the leader's in group by emphasizing similarity in attitudes.

Notes:

In the Situational Leadership Model approach, behavioral scientists identify four leadership styles including delegating, participating, selling, and telling. The delegating leader lets the group members decide what to do themselves, the participating leader asks the members what to do but makes the final decisions himself/herself, the selling leader makes the decision but explains the reasons, and the telling leader makes the decision and tells the group what to do. Each style is appropriate to a specific situation. Delegating works best where followers are willing to do the job and know how to go about it. Participating works best when followers are able to do the job but require emotional support. Selling works best where followers are neither willing nor able to do the job. Telling works best where followers are willing to do the job but don't know how to do it.

Notes:

Transformational Leadership Behavior. Transformational leaders encourage and obtain performance beyond expectations by formulating visions and then inspiring their subordinates to pursue them. They cultivate employee acceptance and commitment to their visions. Transformational leaders provide frames, or the big picture, so subordinates can relate their individual activities to the work as a whole. Transformational leaders are charismatic, inspirational, considerate of individuals, and use intellectual stimulation to encourage employees to approach old, familiar problems in new ways. Examples of transformational leaders include Margaret Thatcher, Jack Welch, and Lee Iacocca.

Notes:

Are There Gender Differences in Leadership Style? Studies suggest few measurable differences in the leader behaviors that women and men managers use on the job. Women managers have been found to be more achievement oriented, while males have been found to be more candid with co-workers. One gender difference was that women were more understanding. Women have been found to perform at about the same level as men. Women often score higher on measures of patience, relationship development, social sensitivity, and communication.

Notes:

Being a Leader

Thinking Like a Leader. Thinking like a leader requires applying the three-step model presented earlier, remembering that leading requires knowledge of matters other than leadership theories, e.g., culture, motivation, groups, conflict, and change, and influencing followers to move toward goals.

Notes:

Develop Your Judgment. Leaders can improve their decision-making ability by increasing their knowledge, debiasing their judgment, being creative, using intuition, not overstressing the finality of decisions, and making sure the timing of a decision is right.

Notes:

Build on Your Other Leadership Traits. Leaders can use good judgment, exhibit self-confidence, and improve their knowledge of the business to improve their effectiveness.

Notes:

Are You Ready to be a Leader? Students can take this self-assessment exercise to give them a feel for their readiness and inclination to assume a leadership role.

Notes:

Rating a Manager's Power. This self-assessment exercise can give students an idea of the steps they can take to enhance their power.

Notes:

Make Sure Your Followers Share Your Vision. The leader should know the objectives of people in the organization and seek to fulfill them. A leader is like the driver of a bus, they should take the organization where it wants to go.

Notes:

<u>**Adapt Your Leadership Style and Actions to the Situation**</u>.

Notes:

<u>**Substitutes for Leadership: User Your Other Management Skills to Help You Lead**</u>.

Notes:

Summary

Notes:

Case: The Ultimate Entrepreneur

1. Matsushita is the classic example of a successful business leader. He created a vision and was able to motivate people to follow him in working toward achieving the mission and objectives of the organization. He obviously was willing to take risks, and he created an environment that fostered growth and opportunity for thousands of employees.

2. In all likelihood, Matsushita s leadership style probably made a difference in the culture of the organization rather than vice-versa. Also, as the firm grew to such extraordinary size, Matsushita had to be an effective delegation while exercising a reasonable amount of control to ensure that objectives were achieved. He had to hire the right people to help grow the business and organize tasks effectively for accomplishment.

3. Important lessons of leadership include treating people as if they were a part of your family, emphasizing customer needs, being accountable for one s own work, and a social responsibility to use resources wisely so as not to fail and commit a crime against society. Also, Matsushita emphasizes that life is growth--as a human being, as a business person, and as a leader.

4. It would be a great opportunity to work in an organization that has grown rapidly and provided jobs for thousands of people. One could learn a lot about leadership and management by following the principles set forth by Matsushita.

5. Matsushita's philosophy and his emphasis on growth as a person as well as in business and leadership made him a leader that many would want to follow. He probably was an example of Chester Barnard s concepts about leader authority being accepted by subordinates.

You be the Consultant: Michael Dorf Leads KnitMedia

1. Dorf appears to be a Transformational Leader because he has charisma, he is inspirational, he encourages employees to self-actualize, and he uses intellectual stimulation to encourage employees to

approach old problems in new ways. He has created a vision for KnitMedia, and he appears to engage in framing to keep people focused on the big picture.

2. Dorf's vision is one of a successful global enterprise in the alternative music and entertainment industry. He will need to delegate effectively and follow up to ensure that objectives are achieved as the organization grows. He will need to let go of his hands-on approach to the hiring of employees and problem solving in the organization.

End of Chapter Questions Chapter 10

Multiple Choice

1. ________________ means influencing others to work willingly toward achieving objectives.

a. Staffing
b. Administering
c. Coaching
d. Leadership
e. Persuasion

Answer: (d) Difficulty: (2)

2. Leadership is an easy concept to define but a difficult one to study and understand. An appropriate definition of leadership would be one which means (pick the most correct and general response available) ________________

a. having others follow you.
b. providing direction for others.
c. influencing others to work willingly toward achieving objectives.
d. accomplishing an objective.
e. having someone follow your commands and being willing to sacrifice to get results.

Answer: (c) Difficulty: (3)

3. What makes leadership a challenging concept to master is the need to understand a variety of leadership theories and to translate these theories into leadership skills through application and practice. All of the following are areas that must be mastered to be a successful leader EXCEPT (pick the least likely general skill or mastery level) ________________

a. how to think like a leader.
b. understanding the foundations of leadership.
c. how to provide a vision.
d. how to act like a leader.
e. how to be forceful with subordinates.

Answer: (e) Difficulty: (2)

4. Two foundations or prerequisites of leadership are _____________ and the right mix of leadership traits.

a. force of will
b. power
c. cunning
d. experience
e. knowledge

Answer: (b) Difficulty: (3)

5. As a leader, a manager must simultaneously fill many roles. All of the following are examples of roles that the leader might be asked to fill <u>EXCEPT</u> (pick the least likely) _____________

a. interacting with bosses.
b. interacting with peers.
c. motivating groups.
d. motivating bosses.
e. managing groups and teams of subordinates.

Answer: (d) Difficulty: (2)

6. As a leader, a manager must simultaneously fill many roles. In addition to leadership skills, the leader must also master other skills in order to perform required roles. All of the following are examples of skills that the leader must master <u>EXCEPT</u> (pick the least likely) ___________

a. culture building.
b. dominance.
c. communication.
d. team building.
e. change leadership.

Answer: (b) Difficulty: (1)

7. When one analyzes the relationship between managing and leading, the conclusion is that leading is (pick the most correct and important) _______________

a. the distinctly behavioral and influencing part of what managers do.
b. the number one skill of managers.
c. really the only real skill that managers need to be successful.
d. the only managerial skill that cannot be learned.
e. a skill that anyone can acquire if they work and try hard.

Answer: (a) Difficulty: (2)

8. ____________________ allow(s) the leader to look at a situation and dig out the underlying assumptions and values that are motivating subordinates, and to evaluate the evidence and think through how to apply what the leader knows about leadership theory to solve the problem.

a. The Wiseman model
b. Critical mass assumptions
c. Critical leadership thinking skills
d. Behavioral mechanics
e. Trait-dominance theory

Answer: (c) Difficulty: (2)

9. Critical leadership thinking skills allow the leader to do all of the following EXCEPT (pick the least likely general skill) ____________________

a. review the leadership situation.
b. identify what is happening.
c. account for what is happening.
d. decide on the leadership actions you will take.
e. make a profit from your actions.

Answer: (e) Difficulty: (2)

10. If a leader is correctly using critical leadership thinking skills to really think like a leader, all of the following steps will be followed to analyze problem situations EXCEPT ____________

a. identifying issues.
b. learning how to control power.
c. analyzing the sequence of cause and effect.
d. synthesizing information.
e. selecting a plan for leading.

Answer: (b) Difficulty: (2)

11. If a leader is correctly using critical leadership thinking skills to really think like a leader, the first step the leader should take is to ________________

a. identify what is happening.
b. learn how to control power.
c. account for what is happening.
d. synthesize information.
e. decide on the leadership actions to be taken.

Answer: (a) Difficulty: (2)

12. If a leader is correctly using critical leadership thinking skills to really think like a leader, the first step the leader should take to correct the following situation would be to ______________________. The situation is "the employees here refuse to divulge bad news to their supervisors."

a. identify what is happening.
b. learn how to control power.
c. account for what is happening.
d. synthesize information.
e. decide on the leadership actions to be taken.

Answer: (a) Difficulty: (2)

13. The second step that leaders generally apply or follow when using critical leadership thinking skills is to ____________________. This means that the leader should understand which theory or theories best explain what is happening. This might be done by linking a leadership or a behavioral concept or theory to the issue.

a. identify what is happening.
b. learn how to control power.
c. account for what is happening.
d. synthesize information.
e. decide on the leadership actions to be taken.

Answer: (c) Difficulty: (3)

14. Consider the following situation: "Employees here refuse to divulge bad news because when they do they are punished by having the supervisor yell at them." In the critical leadership thinking skills model, which of the following steps is most closely associated with the above scenario?

a. identify issues
b. learn how to control power
c. analyze the sequence of cause and effect
d. synthesizing information
e. select a planning for leading

Answer: (c) Difficulty: (2)

15. Consider the following situation: "The employees here refuse to divulge bad news because when they do they are punished by having the supervisor yell at them." Which of the following solutions would be the best solution to this problem? Use critical leadership thinking skills. (Pick the best available answer.)

a. Reassess the personality-based qualifications of the existing supervisors, to transfer or dismiss potential hotheads.
b. Don't worry about the situation. The employees are too sensitive.
c. Train employees on how to not make the supervisors mad.
d. Teach employees how to fix bad situations on their own instead of complaining about the situation. In other words, be proactive.
e. Place a suggestion box at the easy disposal of the employees.

Answer: (a) Difficulty: (2)

16. Consider the following situation: "The employees here refuse to divulge bad news because when they do they are punished by having the supervisor yell at them." All of the following solutions or steps would be possible or potential solutions to this problem EXCEPT (pick the least likely solution or step) ________________

a. reassessing the personality-based qualifications of the existing supervisors, to transfer or dismiss potential hotheads.
b. presenting this data to top managers to encourage them to engage in a training program aimed at making them more open to getting bad news and criticism.
c. administering an attitude survey to all departmental employees to develop data that describe the current feelings about the blame-oriented atmosphere.
d. presenting this data to top managers to see who needs to be immediately fired.
e. developing a system by which quality control issues (or potential bad news) can be discussed in an open format.

Answer: (d) Difficulty: (3)

17. Psychologists Shelley Kirkpatrick and Edwin Locke believe that two important foundation components are prerequisites for having the potential to lead. These components are best described as being ________________

a. power and persuasion.
b. power and personal traits.
c. personal traits and force of will.
d. wealth and power.
e. example and opportunity.

Answer: (b) Difficulty: (3)

18. A leader without power is not really a leader at all. Therefore, to lead one must have power. All of the following are sources of power for leaders EXCEPT (pick the least likely) ______________

a. position power.
b. authority power.
c. expert power.
d. referent power.
e. wealth power.

Answer: (e) Difficulty: (2)

19. To be an effective leader, the leader must have power. ________________ is illustrated by the example of the sales manager or president who has formal authority attached to him/her. Followers are generally expected to respect these leaders as part and parcel of keeping their jobs.

a. Position power
b. Authority power
c. Expert power
d. Referent power
e. Wealth power

Answer: (a) Difficulty: (2)

20. To be an effective leader, the leader must have power. ________________ is illustrated by the example of the leader who can reward employees who do well or coerce or punish those who don't do well (for instance, by dismissing them or sending them home for the day).

a. Position power
b. Authority power
c. Expert power
d. Referent power
e. Wealth power

Answer: (b) Difficulty: (3)

21. To be an effective leader, the leader must have power. ________________ is illustrated by the example of the leader who is respected because of his/her expertise in some area. This expertise is associated with leading (as in the case of the head of a research lab).

a. Position power
b. Authority power
c. Expert power
d. Referent power
e. Wealth power

Answer: (c) Difficulty: (1)

22. If a leader has personal magnetism so that followers follow because of charisma, then the leader is said to have ________________

a. position power.
b. authority power.
c. expert power.
d. referent power.
e. wealth power.

Answer: (d) Difficulty: (2)

23. Identify the source of the following famous quote about leadership: "One ought to be both feared and loved, but as it is difficult for the two to go together, it is much safer to be feared than loved....."

a. Marco Polo, 13th century
b. Henry Ford, 20th century
c. Frederick W. Taylor, 20th century
d. Niccolo Machiavelli, 16th century
e. Chester Barnard, 20th century

Answer: (d) Difficulty: (3)

24. Identify the source of the following thought about leadership: Managers are essentially powerless unless their followers grant them the authority to lead.

a. Marco Polo, 13th century
b. Henry Ford, 20th century
c. Frederick W. Taylor, 20th century
d. Niccolo Machiavelli, 16th century
e. Chester Barnard, 20th century

Answer: (e) Difficulty: (3)

25. Leaders should consider that influencing their people to get their jobs done by relying heavily on their own formal authority or fear is probably a much more imprudent tactic today than it would have been in the 1980s because of ______________

a. court legislation.
b. sexual harassment litigation.
c. downsized, flattened, and empowered organizations.
d. leadership being a vanishing skill.
e. increasing internationalism.

Answer: (c) Difficulty: (2)

26. To be a leader, a manager needs the personality traits to do the job. Identifying what exactly those traits are is the aim of the ______________ theory of leadership.

a. trait
b. personality
c. behavioral
d. autocratic
e. participative

Answer: (a) Difficulty: (1)

27. The idea that leaders are characterized by certain traits was initially inspired by a ______________ concept of leadership.

a. "will to win"
b. "great ego"
c. "to the victor go the spoils"
d. "may the strongest survive"
e. "great man"

Answer: (e) Difficulty: (3)

28. The ______________ concept of leadership would hold that people like General Norman Schwarzkopf and California Senator Diane Feinstein are great leaders because they were born with certain definable personality traits.

a. "will to win"
b. "great ego"
c. "to the victor go the spoils"
d. "may the strongest survive"
e. "great man"

Answer: (e) Difficulty: (3)

29. There is evidence that indicates that there are certain core traits which significantly contribute to business leaders' success. All of the following would be part of that list EXCEPT ________

a. frugality.
b. drive.
c. the desire to lead.
d. honesty/integrity.
e. cognitive ability.

Answer: (a) Difficulty: (2)

30. There is evidence that indicates that there are certain core traits which significantly contribute to business leaders' success. Which of the following leadership traits would be the basis for the fact that leaders are generally action-oriented people with a relatively high desire for achievement, which means they get satisfaction from successfully completing challenging tasks and attaining standards of excellence?

a. Knowledge of business
b. Drive
c. The desire to lead
d. Honesty/integrity
e. Cognitive ability

Answer: (b) Difficulty: (2)

31. There is evidence that indicates there are certain core traits which significantly contribute to business leaders' success. Which of the following leadership traits would be the basis for the fact that leaders are generally the ones in an organization who must pick the right direction and then put into place the mechanisms required to get there? Consider the fact that a leader's intelligence and the subordinates' perception of his or her intelligence are generally highly rated leadership traits.

a. Knowledge of business
b. Drive
c. The desire to lead
d. Honesty/integrity
e. Cognitive ability

Answer: (e) Difficulty: (2)

32. Power and the requisite leadership traits are not sufficient for successful leadership. They are only a foundation. To go beyond this potential, Kirkpatrick and Locke believe that the leader must also have the skill to ____________________

a. always make correct decisions.
b. earn profits.
c. withstand criticism.
d. provide a vision and then engage in behaviors required to implement that vision.
e. impose a will on those who are sometimes unwilling to accept direction and purpose.

Answer: (d) Difficulty: (2)

33. In business administration (as in other areas), it is often required that a leader ________________ or a way out of the wilderness of problems. (Pick the best general answer.)

a. provide direction
b. provide an angle
c. create an edge
d. stifle competition
e. endure personal pain

Answer: (a) Difficulty: (1)

34. A ______________ is a general statement of the organization's intended direction that evokes positive emotional feelings in organization members.

a. position
b. goal
c. vision
d. course
e. plan

Answer: (c) Difficulty: (2)

35. The firm's ________________ defines and operationalizes the top manager's vision.

a. position
b. goal
c. mission statement
d. course
e. plan

Answer: (c) Difficulty: (2)

36. A firm's ________________ serves to communicate "who we are, what we do, and where we're headed."

a. position
b. goal
c. mission statement
d. course
e. plan

Answer: (c) Difficulty: (2)

37. "...to provide any customer a means of moving people and things up, down, and side-ward over short distances..." is an example of a(n) ____________ devised by Otis Elevator Company.

a. position
b. goal
c. mission statement
d. course
e. plan

Answer: (c) Difficulty: (2)

38. ____________ are specific results a leader wants a group to achieve.

a. Objectives
b. Goals
c. Missions
d. Requirements
e. Plans

Answer: (a) Difficulty: (1)

39. ____________, the process of establishing objectives and courses of action, prior to taking action, is the first of the manager's functions.

a. Leading
b. Evaluating
c. Planning
d. Executing
e. Controlling

Answer: (c) Difficulty: (1)

40. Leaders with the power and personal traits required to be effective can provide leadership by engaging in four sets of activities. All of the following are considered to be part of that activity set EXCEPT (pick the least critical) ________________

a. providing a vision.
b. thinking like a leader.
c. using the right leadership style.
d. having technical expertise.
e. using the OB leadership skills (such as influencing culture and motivating).

Answer: (d) Difficulty: (2)

41. As leaders become active in the leadership process, they must eventually choose a leadership style that will be employed in the leadership of subordinates. All of the following are generally accepted leadership styles EXCEPT ________________

a. employee oriented.
b. fear oriented.
c. participative.
d. job centered.
e. autocratic.

Answer: (b) Difficulty: (2)

42. Research has shown that leaders accomplish or perform two basic functions. First, they accomplish the task. Second, they ________________

a. satisfy their superiors.
b. satisfy the needs of the stockholders.
c. satisfy the legal constraints imposed on them.
d. satisfy the needs of group members.
e. satisfy the needs of international influences.

Answer: (d) Difficulty: (3)

43. Generally speaking, the functions of a task-oriented leader are to ____________

a. reduce tension and boost morale.
b. earn a profit.
c. clarify the jobs to be done and force people to focus on their jobs.
d. provide innovative solutions.
e. lead by working hard.

Answer: (c) Difficulty: (2)

44. The role of the social or people-oriented leader is to ________________

a. reduce tension and boost morale.
b. earn a profit.
c. clarify the jobs to be done and force people to focus on their jobs.
d. provide innovative solutions.
e. lead by working hard.

Answer: (a) Difficulty: (2)

45. Research was done at Ohio State University to develop a Leader Behavior Description Questionnaire (LBDQ). The two leadership factors the questionnaire measures have become synonymous with what experts call The Ohio State Dimensions of Leadership. The two factors are best described as being ________________

a. ethical and social structure.
b. consideration and initiating structure.
c. restrictive and behavioral structure.
d. enforcement and reward structure.
e. corrective and instructive structure.

Answer: (b) Difficulty: (3)

46. Research was done at Ohio State University to develop a Leader Behavior Description Questionnaire (LBDQ). The two leadership factors the questionnaire measures have become synonymous with what experts call The Ohio State Dimensions of Leadership. The factor that says "leader behavior is indicative of mutual trust, friendship, support, respect and warmth" is called ________________

a. ethics.
b. restriction.
c. enforcement.
d. instructive.
e. consideration.

Answer: (e) Difficulty: (3)

47. Research was done at Ohio State University to develop a Leader Behavior Description Questionnaire (LBDQ). The two leadership factors the questionnaire measures have become synonymous with what experts call The Ohio State Dimensions of Leadership. The factor that says "leader behavior by which the person organizes the work to be done and defines relationships or roles, the channels of communication, and ways of getting jobs done" is called ________________

a. social structure.
b. initiating structure.
c. behavioral structure.
d. reward structure.
e. instructive structure.

Answer: (b) Difficulty: (3)

48. A(n) ______________________ leader solves the problem and makes the decision himself or herself, using information available at the time.

a. participative
b. dictator
c. autocratic
d. democratic
e. publican

Answer: (c) Difficulty: (2)

49. A(n) ________________ leader shares the problem with his or her subordinates as a group, and together they generate and evaluate alternatives and attempt to reach consensus on a solution.

a. participative
b. dictator
c. autocratic
d. intuitive
e. publican

Answer: (a) Difficulty: (2)

50. If the captain of a ship is faced with the ship being in a condition where it is sinking, the best leadership style to adopt is probably the ________________ style.

a. participative
b. directional
c. autocratic
d. intuitive
e. publican

Answer: (c) Difficulty: (2)

51. If a leader wants members of the group to sincerely "buy into" the decision, the best leadership style to adopt is probably the ______________________ style.

a. participative
b. directional
c. autocratic
d. intuitive
e. publican

Answer: (a) Difficulty: (2)

52. At the same time researchers at Ohio State were developing their Leader Behavior Description Questionnaire for measuring initiating structure and consideration, a similar series of programs was beginning at the University of Michigan. The two leadership theories that emerged from the University of Michigan studies were characterized as being ______________

a. production-centered/employee-centered styles and close/general styles.
b. production-centered/employee-centered styles and contingency theory.
c. contingency theory and path-goal theory.
d. leader-member exchange (LMX) and path-goal theory.
e. situational leadership model and transformational theory.

Answer: (a) Difficulty: (3)

53. ______________ leaders focus on the individuality and personality needs of their employees and emphasize building good interpersonal relationships.

a. Employee-centered
b. Production-centered
c. Contingency-theory
d. Leader-member exchange (LMX)
e. Situational

Answer: (a) Difficulty: (3)

54. __________________ leaders focus on production and the job's technical aspects.

a. Employee-centered
b. Production-centered
c. Contingency-theory
d. Leader-member exchange (LMX)
e. Situational

Answer: (b) Difficulty: (1)

55. Psychologist Fred Fielder is given credit with the development of the __________ style of leadership. Under this idea, the conclusion is that the appropriateness of the leadership style is "contingent upon the favorableness of the group-task situation."

a. employee-centered
b. production-centered
c. contingency-theory
d. leader-member exchange (LMX)
e. situational

Answer: (c) Difficulty: (2)

56. The ______________ leadership theory is based on the expectancy theory of motivation. The expectancy theory states that whether a person will be motivated depends on two things: whether the person believes he or she has the ability to accomplish a task, and his or her desire to do so.

a. employee-centered
b. production-centered
c. contingency theory
d. leader-member exchange (LMX)
e. path-goal

Answer: (e) Difficulty: (2)

57. Under which of the following theories of leadership does the style a leader uses depend on the situation, so that leaders must be flexible and adopt the style that is required?

a. Employee-centered leadership theory
b. Production-centered leadership theory
c. Contingency theory leadership theory
d. Leader-member exchange (LMX) leadership theory
e. Path-goal leadership theory

Answer: (e) Difficulty: (3)

58. Although a leader may have one prevailing style of leadership, the ____________________ theory says that (since most leaders don't treat all their subordinates the same way) leaders may use different styles with different members of the same work group.

a. employee-centered
b. production-centered
c. contingency theory
d. leader-member exchange (LMX)
e. situational

Answer: (d) Difficulty: (3)

59. Which of the following theories suggests that leaders tend to divide their subordinates into an "in" group and an "out" group?

a. Employee-centered
b. Production-centered
c. Contingency-theory
d. Leader-member exchange (LMX)
e. Situational

Answer: (d) Difficulty: (3)

60. Behavioral scientists have developed what they call a situational leadership model to describe how the leader should adapt his or her style to the task. All of the following are considered to be leadership styles appropriate to this model EXCEPT ____________________

a. demanding.
b. delegating.
c. participating.
d. selling.
e. telling.

Answer: (a) Difficulty: (2)

61. Under the situational leadership model, the _____________ leader lets the members of the group decide what to do themselves.

a. demanding
b. delegating
c. participating
d. selling
e. telling

Answer: (b) Difficulty: (1)

62. Under the situational leadership model, the _______________ leader makes the decision himself or herself but explains the reasons.

a. demanding
b. delegating
c. participating
d. selling
e. telling

Answer: (d) Difficulty: (2)

63. According to the situational leadership model, which of the following styles is most appropriate when the followers are willing to do the job and know how to go about doing it?

a. Demanding
b. Delegating
c. Participating
d. Selling
e. Telling

Answer: (b) Difficulty: (2)

64. According to the situational leadership model, which of the following styles is most appropriate when the followers are able to do the job but require emotional support?

a. demanding
b. delegating
c. participating
d. selling
e. telling

Answer: (c) Difficulty: (3)

65. According to the situational leadership model, which of the following styles is most appropriate when the followers are neither willing nor able to the do the job?

a. Demanding
b. Delegating
c. Participating
d. Selling
e. Telling

Answer: (d) Difficulty: (3)

66. According to the situational leadership model, which of the following styles is most appropriate when the followers are willing to do the job but don't know how to do it?

a. Demanding
b. Delegating
c. Participating
d. Selling
e. Telling

Answer: (e) Difficulty: (2)

67. According to material written by James McGregor Burns, ________________ behaviors are "largely oriented toward accomplishing the tasks at hand and at maintaining good relations with those working with the leader (by exchanging promises of rewards for performance)."

a. transactional
b. transformational
c. delegating
d. demanding
e. transgressing

Answer: (a) Difficulty: (2)

68. According to material written by James McGregor Burns, the __________ style of leadership refers to "the process of influencing major changes in the attitudes and assumptions of organization members and building commitment for the organization's mission, objectives, and strategies."

a. transactional
b. transformational
c. delegating
d. demanding
e. transgressing

Answer: (b) Difficulty: (2)

69. Transformational leaders have all the following characteristics EXCEPT __________

a. They are charismatic.
b. They are Christian.
c. They are inspirational.
d. They consider the individual.
e. They use intellectual stimulation to encourage.

Answer: (b) Difficulty: (1)

70. If an employee made the statement (about his/her leader) "I am ready to trust him or her to overcome any obstacle," he/she is mostly likely describing which of the following characteristics of transformational leaders?

a. They are charismatic.
b. They are Christian.
c. They are inspirational.
d. They consider the individual.
e. They use intellectual stimulation to encourage.

Answer: (a) Difficulty: (2)

71. Being a leader means taking the steps required to boost your effectiveness at filling the leader's role. Based on research, all of the following steps are suggested for individuals to improve the chances that, in a leadership situation, they will be a leader EXCEPT __________

a. Command attention by being tough.
b. Think like a leader.
c. Develop your judgment.
d. Make sure your followers share your vision.
e. Adapt your leadership style and actions to the situation.

Answer: (a) Difficulty: (2)

72. Research indicates that there are management style differences between men and women. These differences, however, do not explain why women are not as successful at getting promoted to top management jobs. The best explanation for this situation is ______________

a. "the glass bottomed boat."
b. Lazes-faire leadership.
c. women are not as charismatic.
d. "the glass ceiling."
e. women managers lack vision.

Answer: (d) Difficulty (1)

73. Research indicates that there are management style differences between men and women. These differences, however, do not explain why women are not as successful at getting promoted to top management jobs. All of the following statements are TRUE EXCEPT ______________

a. There are inaccurate stereotypes of women.
b. Women are more understanding leaders.
c. Women posses more patience as a leader.
d. Women managers do not meet performance standards.
e. Women are cooperative and communicative managers.

Answer: (d) Difficulty: (2)

74. Effective leadership requires that one "think like a leader." Which one of the following statements about leadership attitudes is FALSE?

a. A leader must identify what is happening and formulate a leadership response.
b. An effective leader understands that there is probably more than one successful approach to any problem.
c. To succeed a leader must employ all their management skills to motivate their followers to goal directed behavior
d. There is more than one tool in the leadership 'toolbox" that may be used to attain desired objectives
e. Really good leadership style is an inborn characteristic. Successful leaders rely on their "gut feelings" to know what to do.

Answer: (e) Difficulty (2)

75. Possessing leadership traits gives someone the potential ob being a successful leader. One way of enhancing leadership skills is to develop your judgement. All of the following are examples of this strategy EXCEPT ______________

a. Increase your knowledge
b. Remove bias from your decision process
c. Ensure that the timing of your decisions is correct
d. Don't made a decision and then refuse to change it – remain flexible
e. Ignore intuition. Good decisions must be supported by thorough analysis.

Answer: (e) Difficulty: (2)

76. Which of the following is a good leadership trait?

a. Ability to enjoy the conflict that good leadership creates.
b. Ability to exhibit self confidence.
c. Ability to ignore the consequences of your actions.
d. Anility to make decisions with inadequate information.
e. Ability to make decisions quickly.

Answer: (b) Difficulty (3)

77. A powerless leader cannot really be an effective leader. All of the following are strategies used to increase a leader's power base EXCEPT ________________

a. Choose the right followers.
b. Be well-organized in management style to reduce the need to lead the charge for change
c. Exercise whatever power you may have frequently. People will follow a leader that they fear.
d. Identify issues that are important to your followers. Use these to determine the leadership style that is most likely to succeed.
e. Make sure that your subordinates know and understand your mission, vision, and objectives.

Answer: (c) Difficulty: (2)

True/False -- Short Answer

78. ________________ means influencing others to work willingly toward achieving objectives.

Answer: (Leadership) Difficulty: (2)

79. A leader usually only performs one role—that of being the leader.

Answer: (False) Difficulty: (1)

80. In management theory, leading and managing are inseparable.

Answer: (True) Difficulty: (2)

81. Leading is the distinctly behavioral and "influencing" part of what managers do.

Answer: (True) Difficulty: (1)

82. ____________________ allow the leader to look at a situation and dig out the underlying assumptions and values that are motivating subordinates, and to evaluate the evidence and think through how to apply what the leader knows about leadership theory to solve the problem.

Answer: (Critical leadership thinking skills) Difficulty: (3)

83. The first "critical leadership thinking skill" question to be answered to solve a problem is to account for what is happening.

Answer: (False) Difficulty: (2)

84. When leaders attempt to account for what is happening, they are really going through a process of identifying a cause-effect relationship between what has occurred and why.

Answer: (True) Difficulty: (2)

85. History has shown that there is often only one way to solve a leadership problem; therefore, the successful leader has to know what that way is.

Answer: (False) Difficulty: (2)

86. A leader without power is really not a leader at all.

Answer: (True) Difficulty: (1)

87. If a leader has authority power, he/she has the ability to reward, coerce, or punish.

Answer: (True) Difficulty: (2)

88. If a leader has ____________ power, his/her power is based on the leader's personal magnetism so that the leader's followers will follow just because of personal charisma.

Answer: (referent) Difficulty: (3)

89. Niccolo Machiavelli (from his book `The Prince`) believed that leaders were essentially powerless unless their followers granted them the authority to lead.

Answer: (False) Difficulty: (2)

90. The idea that leaders are characterized by certain traits was initially inspired by a "to the victor goes the spoils" concept of leadership.

Answer: (False) Difficulty: (2)

91. A firm's mission statement defines and operationalizes the top manager's vision.

Answer: (True) Difficulty: (1)

92. The leader must influence his or her subordinates to work toward achieving the stated aims of the organization; however, there is no generally accepted or proven theory describing how leaders actually do this.

Answer: (True) Difficulty: (2)

93. The role of a social or people-oriented leader is to clarify the jobs to be done and force people to focus on their jobs.

Answer: (False) Difficulty: (2)

94. The role of the task-oriented leader is to reduce tension, make the job more pleasant, boost morale, and crystallize and defend the values, attitudes, and beliefs of the group.

Answer: (False) Difficulty: (2)

95. The famous University of Michigan studies on leadership focused on production-centered and employee-centered leadership styles under the direction of Rensis Likert.

Answer: (True) Difficulty: (2)

96. ____________ leadership refers to the "process of influencing major changes in the attitudes and assumptions of organization members and building commitment for the organization's mission, objectives, and strategies."

Answer: (Transformational) Difficulty: (3)

Essay

97. What makes leadership a challenging concept to master is the need to understand a variety of leadership theories and to translate these into leadership skills through application and practice. Define leadership and then (in brief outline form) identify the four major approaches that can be used to study and understand leadership. Be sure to indicate any significant questions that should be answered in each of the approaches you discuss.

Answer:

a. Leadership is influencing others to work willingly toward achieving objectives.

b. The general areas of leadership theories in outline form are:

1) How to think like a leader: How to review a leadership situation in which you find yourself and identify what is happening, account for what is happening, and formulate leader actions.
2) What are the foundations of leadership? The two foundations or prerequisites of leadership are power and the right mix of leadership traits, such as drive and self-confidence.
3) Leaders provide a vision. In other words, they provide a general statement of their organization's intended direction and, often, a "roadmap" for how to get there.
4) Leaders act like leaders. To influence followers to work toward the vision, leaders engage in a number of characteristic leader behaviors, such as providing structure and consideration, being participative when the situation warrants, and adapting their leadership style to the situation.

Difficulty: (2)

98. Being a leader requires more than possessing book knowledge about leadership theories. The leader must have critical leadership thinking skills. Define critical leadership thinking skills. Then, list and briefly identify the three steps that are used to review the leadership situation using these critical leadership thinking skills.

Answer:

a. Critical leadership thinking skills are defined as being the skills that allow the leader to look at a situation and dig out the underlying assumptions and values that are motivating subordinates, and to evaluate the evidence and think through how to apply what the leader knows about leadership theory to solve the problem.

b. Three steps that can be used to review the leadership situation are:

1) Identify what is happening. This first step is to detect issues or areas of interest that compel you as a leader to resolve an organizational problem.
2) Account for what is happening. The main purpose of this step is to use behavioral science theories and concepts to account for why the issues you identified are occurring.
3) Decide on the leadership actions you will take. To do this step, the leader will have to apply all the knowledge that has been gained. The leader must learn to think like a leader.

Difficulty: (2)

99. Two famous universities have taken a lead role in studying leadership. Characterize and summarize the theories and thoughts developed at Ohio State University and the University of Michigan with respect to the study and theory of leadership.

Answer:

a. Ohio State University: Initiating structure and consideration have been two of the most frequently used descriptions of leader behavior. They developed out of a research project launched at Ohio State University. The researchers used the Leader Behavior Description Questionnaire (LBDQ).

   Consideration was defined as being leader behavior indicative of mutual trust, friendship, support, respect, and warmth. Initiating structure was defined as leader behavior by which the person organizes the work to be done and defines relationships or roles, the channels of communication, and ways of getting jobs done.

   Research results concluded (rather inconclusively) that in most situations "considerate leaders will have more satisfied subordinates." In addition, the more structuring the leader was, the more grievances were filed. However, where the leader was also very considerate, leader structure and grievances were not related. Note: Students could also discuss participative and autocratic leadership styles here.

b. The University of Michigan: Two leadership theories emerged from the University of Michigan studies. From the studies of Rensis Likert (and his associates) emerged the employee-centered leadership style and the production-or job-centered leadership style.

   Employee-centered styles focus on the individuality and personality needs of the employees and emphasize building good interpersonal relationships.

   Production-or job-centered styles focus on production and the job's technical aspects.

Research results indicated that supervisors with the best records of performance focus their primary attention on the human aspects of their subordinates' problems and on endeavoring to build effective work groups with high performance goals.

Additional research was done on close and general leadership styles. Close supervision was on one end of the continuum that describes the degree to which a supervisor specifies the roles of subordinates and checks up to see that they comply with these specifications. The laissez-faire leader (on the other hand) keeps a complete "hands-off" policy. The general leader is somewhere in the middle of these extremes. The result of research is that most people do not like to be closely supervised (if they are, morale will suffer).

Difficulty: (3)

100. Briefly, explain the path-goal leadership theory.

Answer:

This leadership theory is based on the expectancy theory of motivation. Expectancy theory states that whether a person will be motivated depends on two things: whether the person believes he or she has the ability to accomplish a task, and his or her desire to do so. Robert J. House (who developed the path-goal theory) says that leaders should increase the personal rewards subordinates receive for attaining goals and should make the path to these goals easier to follow (by clarifying it and reducing roadblocks and pitfalls). Under this theory, the style the leader uses depends on the situation, so that leaders must be flexible and adopt the style that is required. This theory has received minimal support because of the difficulty of measuring concepts such as "path."

Difficulty: (2)

101. Demonstrate the differences between transactional and transformational leadership. Identify the four primary characteristics of transformational leaders.

Answer:

a. The transactional leader is oriented toward behaviors that are largely oriented toward accomplishing the tasks at hand and at maintaining good relations with those working with the leader (by exchanging promises of rewards for performance).

b. The transformational leader is oriented toward behaviors that are oriented toward change. The transformational leadership style refers to the process of influencing major changes in the attitudes and assumptions of organization members and building commitment for the organization's mission, objectives, and strategies.

c. Four characteristics of transformational leaders are that (1) they are charismatic, (2) they are inspirational, (3) they give individual consideration to subordinates to help them self-actualize, and (4) they use intellectual stimulation to encourage employees to approach old and familiar problems in new ways.

Difficulty: (2)

Chapter 11

INFLUENCING ORGANIZATIONAL CULTURE AND SHARED VALUES

INTRODUCTION

Studying this chapter will enable students to define organizational culture and values, give examples of healthy and unhealthy corporate cultures, explain how to create and sustain a corporate culture, discuss the importance of using the corporate culture in managing change, and explain how a manager can change an organization's culture.

LECTURE OUTLINE

ANNOTATED OUTLINE

What's Ahead?

Marvin Runyon is faced with the serious, difficult problem of trying to change the organizational culture in the Postal Service. He must decide how to proceed and how to deal with strong resistance from employees.

What Is Organizational Culture?

Organizational culture includes the characteristic traditions, norms, and values that employees share. Values and norms are basic beliefs about what one should or should not do and what is or is not important. Norms and values guide and channel people's behavior, thus leading and influencing the norms and values they use as behavioral guides.

Notes:

A Closer Look at What Organizational Culture Means. There are several layers of culture in an organization. These include cultural artifacts that are the obvious signs and symbols of corporate culture such as written rules, office layouts, organizational structure, and dress codes. Supporting this layer are patterns of behavior including ceremonial events, written and spoken comments, and actual behavior of managers and employees. These corporate cultural signs and behaviors are a product of the values and beliefs of employees. Management's basic assumptions are at the core of a company's culture.

Notes:

Corporate Culture at Proctor & Gamble. Proctor & Gamble's corporate culture reflects the firm's legendary emphasis on thoroughness, market testing, and ethical behavior. These values are transmitted to new employees through the firm's selection, socialization, and training.

Notes:

Unhealthy Organizational Culture. An unhealthy corporate culture can undermine a firm's effectiveness. Strong, arrogant corporate cultures can develop when managers concentrate on internal administration and politics of the firm rather than on the company's outside customers, stockholders, and competitors.

Notes:

How to Create and Sustain the Corporate Culture

Publish a Core Values Statement. Publishing a core values statement is a logical first step in creating a culture. The values then guide the behavior of all employees.

Notes:

Exercise Leadership. Leaders create and sustain a firm's culture through the actions they take, the comments they make, and the visions they espouse. Founders of organizations are especially important in this regard, e.g., Sam Walton of Wal-Mart.

Notes:

Use Management Practices. Managers can use planning, organizing, leading, and controlling to foster the desired culture. Founders of organizations have a tendency to hire people whose values and beliefs are similar to their own. This helps to perpetuate the culture desired by the founder.

Notes:

Use Signs, Symbols, Stories, Rites, and Ceremonies. It is the leader's responsibility to create strong cultures by shaping norms, instilling beliefs, inculcating values, and generating emotions. Leaders achieve this by establishing management practices that symbolize the firm's core values and by using symbols, stories, rites, and ceremonies.

Notes:

Corporate Culture's Role in Managing Change

Company Culture and the Responsive Organization. Corporate culture affects a company's responsiveness and some values support responsiveness better than others. Values act as guides to help managers delegate more authority and let employees make quick decisions. Normative control means the required efforts of employees are elicited and directed through control of the underlying values, thoughts, beliefs, and feelings that guide their actions. Employee behavior in responsive firms is guided by a strong set of shared values, beliefs, and traditions--a strong culture.

Notes:

Comparing Responsive and Unresponsive Cultures. Researchers Kotter and Heskett concluded, "Only cultures that can help organizations anticipate and adapt to environmental change will be associated with superior performance over long periods of time." They found several differences between adaptive and unadaptive cultures. Adaptive cultures emphasized risk taking, trusting, and a proactive approach to identifying problems and implementing solutions. Unadaptive cultures were bureaucratic where employees were reactive, quick to blame, risk averse, and not very creative. Communications were affected by negative values like turf protection being emphasized over getting the job done.

Notes:

Changing Organizational Cultures

When Cultures Must Be Changed. Often, the difference between successful firms and failures is the ability to change culture fast enough. A firm may be forced to change its culture due to a crisis, e.g., deregulation or the introduction of new technology by a competitor. Turnover in top management or the stage in the firm's life cycle may also create the need for cultural change.

Notes:

Implementing Cultural Change. The most important factor that distinguishes successful cultural change from unsuccessful efforts is competent leadership from top management. Implementing cultural change requires publishing a new formal core values statement and publicizing it widely. Leadership must be able to influence others to work willingly and behave consistently with desired cultural values. The management functions are also used to change culture, e.g., by hiring employees whose values are consistent with the desired culture and using signs, symbols, stories, rites, and ceremonies to emphasize the desirable values and encourage employees to adopt these values as their own.

Notes:

General Electric successfully implemented cultural change in the 1980s when it moved from a sluggish, diverse set of businesses to a lean, cohesive market leader. The firm was driven by the need to make the company more competitive. Jack Welch earned the nickname "Neutron Jack" for his efforts as a change agent. He knew that downsizing the organization was not enough; the culture needed to be changed also. He created a values statement and instituted management practices that epitomized the values. He decentralized authority while fostering teamwork by promoting candor and trust. The results were dramatic in revenue and earnings growth.

Notes:

Summary

Notes:

Case: Here Come the Clowns!

1. The key to Cognex's success appears to be the entrepreneurial culture that has been developed by the founder, Robert Stillman. Humor is used to open communication between managers and employees and build a motivated workforce.

2. Founders play a key role in culture formation and development as they create a vision of where they want the organization to go. In this case, Stillman has used humor to break down barriers between managers and employees and to obtain cooperation and support. He begins the effort in employee orientation to immediately introduce new employees to the culture.

3. Students will have their own opinions about working for Cognex. Some may prefer a more serious approach, but the majority would probably enjoy working in the positive environment. After all, why can't work be enjoyable as long as the firm is accomplishing its mission and objectives by managing assets, including employees, effectively and efficiently and producing one of the fastest growth rates in industry?

4. Firms can learn the importance of communicating cultural values in employee orientation and the significance of top management's actions and behavior in cultural maintenance. Also, perhaps more firms should recognize how humor can break down communications barriers between managers and employees and help create an entrepreneurial culture.

You Be the Consultant: The KnitMedia Culture

1. KnitMedia tries to develop a freewheeling culture that emphasizes innovation, improvisation, and creativity to achieve objectives. Students can examine what they have studied in the case thus far and

specify the underlying values and assumptions. Dorf seems to prefer hiring employees who like to work in an organization where they are given freedom to act but are still held accountable for their actions. He tries to create a culture where employees can be creative and innovative as long as they are accomplishing results.

2. The only major concern is whether Dorf can continue to maintain the same culture as the firm continues growing and employees become further removed from the visionary leader. He will need to find new ways to maintain the same small-company atmosphere as the firm grows.

3. Dorf will need to issue a values statement, ensure that employees are hired whose values and beliefs are consistent with it, and ensure that his managers are acting and communicating in a manner consistent with the desired culture. As the firm grows, he will need to hire a human resource manager to help in these efforts.

End of Chapter Questions for Chapter 11

Multiple Choice

1. The task facing leaders in modern organizations is to inspire and manage change. However, to do so they must also manage (and sometimes change): (pick the best general answer) ____________

b. people in the organization.
c. supervisors.
d. the international environment.
e. the legal environment of the organization.
f. the values and culture of the organization they are leading.

Answer: (e) Difficulty: (2)

2. ____________________ can be defined as the characteristic traditions, norms, and values that employees share.

a. Organizational behavior
b. Organizational traditions
c. Organizational culture
d. Normative behavior
e. Normative culture

Answer: (c) Difficulty: (3)

3. Leaders influence an organization's culture. All of the following are factors that contribute to this influence of an organization's culture EXCEPT (pick the least likely) ________________

a. how the leader acts.
b. how frugal the leader is.
c. by the core values and statements formulated by the leader.
d. by management practices put in place by the leader.
e. by the signs, symbols, and stories encouraged by the leader.

Answer: (b) Difficulty: (1)

4. An organizational culture is defined as the characteristic (pick the best general statement from those listed) ________________

a. trademark and branded labels of the organization.
b product advertising of the organization.
c. traditions, norms, and values the employees share.
d. racial composite of the firm.
e. gender orientation of the organization and its willingness to accept change.

Answer: (c) Difficulty: (2)

5. __________ are basic beliefs about what you should or shouldn't do, and what is or is not important.

a. Principles
b. Traditions
c. Culture adaptations
d. Values and norms
e. Leadership goals

Answer: (d) Difficulty: (2)

6. "Be honest," "Be thrifty," and "Don't be bureaucratic" are basically considered to be ______________ reflecting basic beliefs about what you should or shouldn't do and what is or is not important.

a. values and norms
b. traditions
c. principles
d. leadership statements
e. mandates

Answer: (a) Difficulty: (3)

7. __________________ guide and channel people's behavior. Therefore, leading and influencing people depend in part on influencing these individual guides to personal behavior. (Pick the general answer that best fits the organizational leadership issue.)

a. Values and norms
b. Traditions
c. Principles
d. Leadership statements
e. Mandates

Answer: (a) Difficulty: (3)

8. All of the following are influences on the organization's culture EXCEPT: (pick the least likely influence) ________________

a. the company's core values statement.
b. the employee's pre-existing values.
c. the organization's traditions.
d. the political affiliation of the leader.
e. signs, symbols, stories, rites, and ceremonies encouraged by the leader.

Answer: (d) Difficulty: (2)

9. Whether the leader's influence on culture is positive or negative depends on the ____________

a. employee's perception of modern business practices.
b. employee's perception of the profit mandate.
c. values the leader espouses and on how he or she molds the company's culture.
d. employee's condition of happiness.
e. strength of the competition.

Answer: (c) Difficulty: (2)

10. When thinking generally about culture (using the United States as an example), culture can mean the (pick the best general answer) ____________________

a. intelligence of the average citizen.
b. values the country's citizens share.
c. number of banks the country has.
d. number of professional sports teams the country has.
e. ability to write and understand Latin.

Answer: (b) Difficulty: (2)

11. When thinking generally about culture (using the United States as an example), all of the following can mean "culture" EXCEPT: (pick the least likely answer) ______________

a. values the country's citizens share.
b. values another country's citizens share.
c. physical aspects of the country's society (such as museums, libraries, or art galleries).
d. the characteristic ways the people of that country behave.
e. the artistic aspects of the country (such as art, music, theater).

Answer: (b) Difficulty: (3)

12. To many, culture means the physical aspects of a country's society. All of the following would be examples of this meaning EXCEPT ___________________

a. a country's art.
b. a country's music.
c. a country's theater.
d. a country's bank system.
e. a country's literature.

Answer: (d) Difficulty: (1)

13. If a cultural indication was given by a country's emphasis on "hard work" or "democracy," then the aspect of culture that would be under examination would be ___________________

a. values the country's citizens share.
b. values another country's citizens share.
c. physical aspects of the country's society.
d. the characteristic ways the people of that country behave.
e. the artistic aspects of the country.

Answer: (a) Difficulty: (2)

14. ___________________ are the obvious signs of corporate culture, such as written rules, office layouts, organizational structure, and dress codes.

a. Cultural artifacts
b. Leadership artifacts
c. Patterns of behavior
d. Values and beliefs
e. Basic assumptions

Answer: (a) Difficulty: (2)

15. ___________________ include the company's ceremonial events, written and spoken comments, and the actual behaviors that the firm's managers and other employees engage in (such as hiding information, politicking, or expressing honest concern when a colleague needs assistance).

a. Cultural artifacts
b. Leadership artifacts
c. Patterns of behavior
d. Values and beliefs
e. Basic assumptions

Answer: (c) Difficulty: (3)

16. According to Fred Smith, the founder and chairman of Federal Express, "Many firms say they believe in respecting their employees and putting their people first." However, if the leader insists on time clocks, routinely downsizing, and reducing employee benefits, the real __________________ of the leader come out. (Pick the best available statement.)

a. cultural artifacts
b. leadership artifacts
c. patterns of behavior
d. values and beliefs
e. basic assumptions

Answer: (d) Difficulty: (2)

17. Which of the following layers are considered by James Hunt to be at the core of an organization's culture? In other words, other actions stem from this layer.

a. Cultural artifacts
b. Leadership artifacts
c. Patterns of behavior
d. Values and beliefs
e. Basic assumptions

Answer: (e) Difficulty: (2)

18. If the manager of a retail store said, "The employees of this store are basically dishonest and must be watched closely," which of the following layers organizational culture would the manager be speaking from?

a. Cultural artifacts
b. Leadership artifacts
c. Patterns of behavior
d. Values and beliefs
e. Basic assumptions

Answer: (e) Difficulty: (3)

19. With respect to organizational culture, it has been shown that assumptions mold values, which in turn mold behavior and the physical manifestations of a person's behavior (like policies and rules). All of the following are considered to be basic types of assumptions EXCEPT __________________

a. an organization's relationship to environment.
b. the nature of reality, truth, time, and space.
c. the nature of human nature.
d. the wealth of capitalism.
e. the nature of human activity.

Answer: (d) Difficulty: (3)

20. With respect to organizational culture, it has been shown that assumptions mold values, which in turn mold behavior and the physical manifestations of a person's behavior (like policies and rules). The assumption called ________________ examines the extent to which key organizational members see this relationship as one of dominance, submission, harmonizing, finding an appropriate niche, and the like.

a. an organization's relationship to environment
b. the nature of reality, truth, time, and space
c. the nature of human nature
d. the nature of human relationships
e. the nature of human activity

Answer: (a) Difficulty: (3)

21. With respect to organizational culture, it has been shown that assumptions mold values, which in turn mold behavior and the physical manifestations of a person's behavior (like policies and rules). The assumption called ________________ examines what it means to be "human," and what attributes are assumed be to intrinsic or ultimate. Additional questions would be: Is human nature evil, good, or neutral? Is Theory X or Y appropriate?

a. an organization's relationship to environment
b. the nature of reality, truth, time, and space
c. the nature of human nature
d. the nature of human relationships
e. the nature of human activity

Answer: (c) Difficulty: (2)

22. At Proctor & Gamble, there is an emphasis on orderly growth. This manifests itself in "tremendous conformity." A new recruit soon learns to say "we" instead of "I." This would be an illustration of which of the following corporate culture concepts?

a. A corporate truth
b. A corporate value
c. A corporate rule
d. Corporate discipline
e. Corporate administration

Answer: (b) Difficulty: (2)

23. Corporate values are present in all of major organizations. All of the following are examples of corporate values <u>EXCEPT</u> (pick the least likely general value) ________________

a. corporate conformity.
b. testing through research and development.
c. an emphasis on written rather than verbal communications.
d. secrecy.
e. price discounting.

Answer: (e) Difficulty: (2)

24. A company's culture can be a force for good or ill. Research by Kotter and Heskett has discovered underlying causes or origins of unhealthy cultures in organizations. All of the following are sources of an unhealthy culture EXCEPT ________________

a. The firm hires, and promotes managers, not leaders.
b. Focus is on internal politics rather than external constituencies.
c. Managers become arrogant.
d. Managers begin to attend more self-evaluation seminars.
e. Managers begin to believe that their idiosyncratic traditions are superior.

Answer: (d) Difficulty: (3)

25. A company's culture can be a force for good or ill. Research by Kotter and Heskett explored the underlying causes or origins of unhealthy cultures in organizations. Which of the following would be considered to be one of these origins?

a. The firm hires and promotes leaders.
b. The firm's primary focus is on external constituencies.
c. Managers become arrogant.
d. Leaders discourage internal politicking and territory development.
e. Mangers are open to new ideas and ways of doing things.

Answer: (c) Difficulty: (2)

26. Consider the following statement: The need to manage success drives management to permit an administrative bureaucracy to develop in which shuffling papers and protecting one's turf become more important than providing and following a visionary direction for the firm. The above statement characterizes which of the following organizational cultures?

a. A leader-driven culture
b. An unhealthy culture
c. A dynamic culture
d. A competitive culture
e. A change culture

Answer: (b) Difficulty: (1)

27. Managers can take concrete steps to create and sustain a healthy corporate culture. These include all of the following steps EXCEPT ________________

a. publishing a competitive manifesto.
b. exercising leadership.
c. using management practices.
d. publishing a core values statement.
e. establishing signs, symbols, and ceremonies.

Answer: (a) Difficulty: (3)

PRENTICE HALL, Upper Saddle River, New Jersey 07458

http://www.prenhall.com

ISBN 0-13-012898-8
99991
9 780130 128980

UNTANGLING ANXIETY

A GUIDE TO RECLAIMING CONTROL & CONFIDENCE

PIPER HARRIS, MA, NPT-C